THE EU–RUSSIAN ENERGY DIALOGUE

The International Political Economy of New Regionalisms Series

The International Political Economy of New Regionalisms Series presents innovative analyses of a range of novel regional relations and institutions. Going beyond established, formal, interstate economic organizations, this essential series provides informed interdisciplinary and international research and debate about myriad heterogeneous intermediate level interactions.

Reflective of its cosmopolitan and creative orientation, this series is developed by an international editorial team of established and emerging scholars in both the South and North. It reinforces ongoing networks of analysts in both academia and think-tanks as well as international agencies concerned with micro-, meso- and macro-level regionalisms.

Recent titles in the series

An East Asian Model for Latin American Success
The New Path
Anil Hira

European Union and New Regionalism
Regional Actors and Global Governance in a Post-Hegemonic Era
Second Edition
Edited by Mario Telò

Regional Integration and Poverty
Edited by Dirk Willem te Velde and the Overseas Development Institute

Redefining the Pacific?
Regionalism Past, Present and Future
Edited by Jenny Bryant-Tokalau and Ian Frazer

The EU–Russian Energy Dialogue

Europe's Future Energy Security

Edited by

PAMI AALTO
University of Tampere

Routledge
Taylor & Francis Group
LONDON AND NEW YORK

First published 2008 by Ashgate Publishing

Published 2016 by Routledge
2 Park Square, Milton Park, Abingdon, Oxon OX14 4RN
711 Third Avenue, New York, NY 10017, USA

Routledge is an imprint of the Taylor & Francis Group, an informa business

British Library Cataloguing in Publication Data
The EU-Russian energy dialogue : Europe's future energy security. - (The international political economy of new regionalisms series)
1. Energy policy - European Union countries 2. European Union countries - Foreign economic relations - Russia (Federation) 3. Russia (Federation) - Foreign economic relations - European Union countries
I. Aalto, Pami, 1972-
333.7'9'094'090511

Library of Congress Cataloging-in-Publication Data
The EU-Russian energy dialogue : Europe's future energy security / edited by Pami Aalto.
p. cm. -- (International political economy of new regionalisms series)
Includes index.
ISBN 978-0-7546-4808-6
1. Energy policy--European Union countries. 2. Energy policy--Russia (Federation) 3. Power resources--European Union countries. I. Aalto, Pami, 1972-

HD9502.E82E88 2007
333.79094--dc22

2007020233

ISBN 9780754648086 (hbk)

Contents

List of Tables

List of Figures

List of Abbreviations

ASPO	Association for the Study of Peak Oil and Gas
BASREC	Baltic Sea Region Energy Co-operation
BEAC	Barents Euro-Arctic Council
CBSS	Council of the Baltic Sea States
CCB	Coalition Clean Baltic
CEE	Central and Eastern Europe
CES	Common Economic Space
CFSP	Common Foreign and Security Policy
CIS	Commonwealth of Independent States
DG TREN	Directorate General on Transport and Energy
EBRD	European Bank for Reconstruction and Development
ECEN	Estonian Council of Environmental NGOs
ECSC	European Coal and Steel Community
ECT	Energy Charter Treaty
EEA	European Economic Area
EEC	European Economic Community
EMU	European Monetary Union
ENP	European Neighbourhood Policy
ESDP	European Security and Defence Policy
EU	European Union
EURATOM	European Atomic Energy Community
EIB	European Investment Bank
FCMA	Friendship, Co-operation and Mutual Assistance treaty
FDI	Foreign direct investment
GATT	General Agreement on Trade and Tariffs
GDP	Gross Domestic Product
GSEO	Group of senior energy officials
HELCOM	Helsinki Commission
IEA	International Energy Agency
IMF	International Monetary Fund
IMO	International Maritime Organisation
LNG	Liquified natural gas
NATO	North Atlantic Treaty Organisation
NCM	Nordic Council of Ministers
NGO	Non-governmental organization
ND	Northern Dimension
NDEP	Northern Dimension Environmental Partnership
NIB	Nordic Investment Bank

OSCE	Organization for Security and Co-operation in Europe
OPEC	Organization for Petroleum Exporting Countries
PCA	Partnership and Co-operation Agreement
PSA	Production-sharing agreement
RSPP	Russian Union of Industrialists and Entrepreneurs
TEN	Trans-European Networks
WTO	World Trade Organization

Contributors

Pami Aalto is acting Jean Monnet Professor in the Department of Political Science and International Relations, University of Tampere, and a member of Eurasia Energy Group, Aleksanteri Institute, University of Helsinki. His most recent books are *Constructing Post-Soviet Geopolitics in Estonia* (2003) and *European Union and the Making of a Wider Northern Europe* (2006).

Eiki Berg is Professor of International Relations in the Department of Political Science, University of Tartu. He is co-editor of *Mapping Borders between Territories, Discourses and Practices* (2003) and special issue of *Journal of Baltic Studies* on 'Baltic Sea Region: Territoriality, Multi-Level Governance and Cross-Border Networks (2002).

Viatcheslav Morozov is Associate Professor in the School of International Relations, St. Petersburg State University. He is co-editor of *Market Democracy in Post-Communist Russia* (2005), and author of numerous articles on Russian foreign and regional policy published for example in *Cooperation and Conflict* and *Journal of Baltic Studies*.

Tatiana Romanova is Associate Professor in the School of International Relations, St. Petersburg State University. She is author of *European Union as an International Actor* (2003, in Russian), as well as numerous articles on energy and economic policy in EU–Russia relations.

Stanislav L. Tkachenko is Vice-Rector at the St. Petersburg State University and Associate Professor in the School of International Relations. He is co-author of *Military Cooperation in the CIS and the Influence of NATO* (2002), and author of two books on Russian financial policy. He has published widely in international compilation works and also in *Communist and Post-Communist Studies*.

Nina Tynkkynen is PhD student in the Department of Regional Studies, University of Tampere. She is finalizing her dissertation on Russia and multilateral environmental cooperation, and has published articles in *Environmental Politics* and in several compilation works.

Kirsten Westphal is Researcher and Lecturer in the Department of Political Science, Justus-Liebig-Universität, Giessen. She is author of *Russische Energiepolitik* (2000) and co-editor of *Conflicts in a Transnational World* (2006) as well as *Focus on EU–Russia Relations* (2005).

Acknowledgements

Preparations for this book were started at a time when energy was only assuming the importance that it now has on the contemporary policy agenda. The initial idea for this book was conceived already in 2003. The author group was assembled and format for the book was agreed during 2004. Our research assistants Helena Koff in Tartu and Kristina Minkova in St. Petersburg compiled valuable material, which for its own part helped us to conduct the bulk of the research during 2004–5. During the research and writing-up process we started to feel very concretely that we are working on something topical. Since then we have been amazed at what speed events in European, European–Russian and wider energy politics unfold, and have done our best to keep abreast with the developments whilst also, importantly, looking towards the future in many contributions in this book, without hopefully forgetting the equally important past events and legacies of energy policy. Many individuals have crucially helped by commenting on our work during the process, and whilst they are too numerous to be all mentioned, we trust to continue the intellectual and policy exchange in the future, also after the publication of this book.

The first public seminar and internal book workshop related to the book were held in connection to the 5th symposium for Russian and East European Studies, arranged by the Finnish Association for Russian and East European Studies, 17–18 March 2005, in Tampere, Finland. Our discussant Esa Kokkonen helped us to gear our policy agenda during that seminar. We were encouraged by the response, intensified our work and discussed the contributions to the book in more detail during an intensive two-day book workshop in the School of International Relations, St. Petersburg State University, 10–11 February 2006. Oksana Grigorieva, Andrey Shadurskiy and Elena Fofanova kindly and effectively acted as our external discussants in that workshop, and Dean Konstantin Khudoley generously provided facilities for our meeting.

We are also grateful for the financial support from the Academy of Finland's 'Russia in Flux' programme (2004–7), project 'New and Old Russia in the Transition Discourses of Finnish–Russian Relations' (project leader Helena Rytövuori-Apunen, no. 1208148/SA). While all contributors have received important support from their own institutions in Finland, Russia, Estonia and Germany, the editor would like to reserve special thanks to the institutions where he has coordinated the work during the course of the editing process: Department of Political Science and International Relations, University of Tampere, where the project was initiated; Department of Political Science, University of Helsinki, where very useful help during the fine-tuning of the book was provided by Robyn Milburn during spring 2006; and Aleksanteri Institute, University of Helsinki, where the major part of the editing was done, and where the editor had all the support he needed and also enjoyed greatly the work within the Institute's Eurasia Energy Group. At the same time the editor would like to thank Kirstin Howgate and Margaret Younger at Ashgate for all help and patience during the editing, and the series editor Timothy Shaw for an enthusiastic attitude and proactive approach towards the project that crucially helped to keep going. No less important are the editor's personal thanks to his family, Freja and Hugo, for a caring and warm environment tolerating all these research efforts, and helping to keep things in perspective in such a rapidly changing field as energy policy.

Introduction

Pami Aalto and Kirsten Westphal

> Oil is such a vital commodity for western Europe that any evidence which suggests some uncertainty or threat over its continuing supply will obviously be investigated very carefully. And when a threat looks real enough, and seems to be unavoidable, every country in Europe will naturally start examining its future energy options – just in case. Above all, answers to a lot of difficult questions will be needed.[1]

Europe Threatened by Energy Problems

The vitality of oil for European countries that Gordon Goodman notes in the quote above has not really changed much since he originally put forth the remark in 1981. Oil simply continues to be vital. But alongside oil there are increasing concerns for the supply of natural gas, as well as for other forms of energy. This means that today, we are speaking of a wide array of questions pertaining to European energy policy. Some of these questions are evident in the energy related problems and crises that have been witnessed during the new millennium.

The considerable oil price hikes since 2001–2 reproduced the European and global concerns for oil prices and supplies that arose with the 1970s oil crises. This time the price hikes were a consequence of several factors: the politicized struggle over the control of the Yukos oil giant in Russia during 2003–5, political unrest in oil producers Nigeria and Venezuela, unusually potent storms disrupting oil drilling and production along the US coast in 2005, and perhaps most importantly, the continuation of the US led war in oil rich Iraq since 2003. Oil prices had already been climbing for a few years from their lows of only 10–15 US dollars a barrel in 1998, but these events inched prices at over 70 US dollars in summer 2006.

At the turn of 2005 and 2006, the vital gas deliveries to the European Union (EU) from Russia, also for a while seemed to be endangered. Around 78 per cent of the gas deliveries to EU territory go through pipelines crossing Ukraine.[2] The

---

1 Gordon T. Goodman, 'Preface', in Gordon T. Goodman, Lars A. Kristoferson and Jack M. Hollander (eds), *The European Transition from Oil: Societal Impacts and Constraints on Energy Policy* (London: Academic Press, 1981), p. vii.

2 This figure includes the Brotherhood, Soyuz and Northern Light pipelines, and the Transbalkans pipeline supplying Romania and Bulgaria. The completion of the Yamal–Europe pipeline through Belarus and Poland in 2005 reduced Ukraine's relative weight as a transit country from the previously even higher figures; see *Russian Analytical Digest*, 6 June 2006, p. 4; see also Aleksanteri Institute Eurasia Energy Group, 'The Energy Dynamic on

state-dominated Russian gas giant Gazprom and Ukraine were wrangling over gas prices for Russia's deliveries to the country, and transit fees for Russian gas on its way to the wider European markets. For a few days Gazprom's customers in western and central Europe reported reduced pressure levels in their gas pipes, sparking the question of the sufficiency of gas reserves, and, ultimately, alternative sources and supply routes.

Electricity supply blackouts emerged in the beginning of the new millennium in locations ranging from California across the Atlantic to Finland in northern Europe, Germany and Russia, due to failures in power supply line systems, dry summer and the consequent reduction in hydropower generated electricity supplies, and heavy snowfall in winter-time. At the same time electricity prices were climbing for instance in the Nordic electricity market Nordpool from 30 euros per megawatt in 2005 to over 50 euros in early 2006. Nuclear energy, which for years had looked politically a too sensitive mode of energy production after the explosion of the Chernobyl reactor in the then Soviet Ukraine in 1986, became a catchword for European energy policy makers again regardless of long-standing popular resistance to it and environmental concerns of the safety of storing nuclear waste.

In short, in the new millennium, energy issues have re-entered the global, and by implication, European policy agenda, and look set to stay there. However, two and half decades after Goodman's remarks on oil shortages in Western Europe, European countries are situated in an entirely different context in which to respond to policy making problems. This follows from the emergence of the EU as Europe's main economic integration machine and an all-European political process that has in many senses rendered Goodman's term 'western European countries' politically incorrect. Efforts of responding to energy policy concerns at the EU level and in that way building a *European energy policy* were evident for example during Austria's EU presidency in the first half of 2006, when an Energy Community secretariat was established in Vienna, seeking to create a joint energy market between the Union and the Balkans. The following Finnish presidency also took up energy as a key item on its agenda, as did Germany in spring 2007. But regardless of this infiltration of the agenda of EU presidencies by energy issues, it is the contention of this volume that most likely, we have not yet seen the real peak of European energy debates.

What we have seen in Europe during the new millennium, is increasing awareness of the centrality of energy for sustaining economic, political and social life, as well as awareness of its scarcity. This is taking place in a situation where none of the environmental concerns related to energy production and transportation have disappeared, but rather have become amplified with global warming associated with the use of hydrocarbons like oil, natural gas and coal, and the spectre of further oil transportation catastrophes in the seas after events such as the large-scale oil spills from the wreckage of the Prestige tanker in the Spanish coast in November 2002. Moreover, the EU has been caught as still struggling to establish itself if not a leading, then at least a coordinating position at the face of its western, eastern, northern and southern member states often in the first place looking to secure their own supplies

---

the Borders of the EU: Belarusian–Russian Relations', policy presentation <www.helsinki.fi/aleksanteri/energy>.

by bilateral agreements with energy producers. We suggest in this volume that greater unity is needed in order to solve what essentially are shared energy problems resulting from the unequal distribution of energy resources across the wider European area, and hence, interdependence between producers and consumers in the energy market. This can effectively be read as a call for a *pan-European energy policy*[3] geared at avoiding the problem of the commons and promoting collectively good solutions.

The Challenge of the EU–Russia Energy Relations

A central policy axis regarding our call for a pan-European energy policy is the EU–Russia energy relationship. In October 2000 the *EU–Russia energy dialogue* was launched to develop it further. In this volume we will focus on its implications for dealing with energy policy problems in Europe and the wider pan-European area.

What we so far know about the EU–Russia energy dialogue comes mostly in the form of empirically informative articles and reports with a descriptive approach, often coupled with a policy analytical ambition. One approach concentrates on EU–Russia *energy diplomacy*, and discusses how it highlights the problems and prospects of economic integration in the Eurasian continent whilst simultaneously giving some new political impetus for EU–Russia relations in the political sphere.[4] *Geopolitics and energy security approaches* situate the EU–Russia energy dialogue into the context of global geopolitics, interstate and interregional competition, and sometimes raise the spectre of proliferating resource conflicts and even wars.[5] In some cases the geopolitical approach is linked to the widespread research on *energy economics and trade*.[6] Energy diplomacy, geopolitics and energy economics and trade are also approaches that characterise energy policy research on the whole.

3 This has also been demanded by the European Commission, 'Green Paper: A European Strategy for Sustainable, Competitive and Secure Energy', Brussels, 8 March 2006, COM(2006) 105 final, p. 16. The 2007 Communication of the EU Commission on 'An Energy Policy for Europe' is much more cautious about the prospects of such an energy community; European Commission, 'An Energy Policy for Europe', COM(2007) 0001 final, Brussels, 10 December 2007.

4 E.g. Debra Johnson, 'EU–Russia Energy Links', in Debra Johnson and Paul Robinson (eds), *Perspectives on EU–Russia Relations* (London: Routledge, 2005); Vladimir Voloshin, 'EU–Russia Energy Dialogue', Russian–European Centre for Economic Policy, 2004.

5 E.g. Amy Myers Jaffe and Robert A. Manning, 'Russia, Energy and the West', *Survival*, 43/2 (summer 2001): 133–52; Susanne Peters, 'Courting Future Resource Conflict: the Shortcomings of Western Response Strategies to New Energy Vulnerabilities', *Energy Exploration and Exploitation*, 21 (2003): 29–60; Gawdat Bahgat, 'Europe's Energy Security: Challenges and Opportunities', *International Affairs*, 82/5 (September 2006): 961–75.

6 E.g. Aad Correlje and Coby van der Linde, 'Energy Supply and Geopolitics: A European Perspective', *Energy Policy*, 34 (March 2006): 532–43; Jonathan Stern, 'The New Security Environment for European Gas: Worsening Geopolitics and Increasing Global Competition for LNG', Oxford Institute for Energy Studies NG 15/2006.

Within this wider context, a further commonplace approach focuses on *energy and the environment*.[7]

We think that energy diplomacy, geopolitical and energy economics approaches have an important place in the field, and in this book, we make conscious overtures towards these research directions. But the first instance where we wish to look for new directions is to delve deeper into the wider ramifications of energy issues as witnessed in the EU–Russia energy dialogue. In this case, these pertain not only to how the dialogue spills over from economic and technical issues to the more explicitly political and often security political realms, as well as to the environmental sector. At issue are also questions on what the EU's and Russia's activities in this sphere tell us about the role of energy in the past, present and future of European integration; and what role energy plays in Russia's economic and political transformation prospects. By opening this line of enquiry that can be called the *political sociology of energy*, we trust to make some new openings into the policy debates on the merits and shortcomings of the energy dialogue, and outline some new ideas for its further development. The second instance where we will probe some new terrain is by covering not only the EU and Russia as monolithically understood entities, but also selected EU member states, various bureaucratic actors within them, and their links with energy companies and other interest groups. This approach can be called the *bureaucratic politics of energy*. Towards the end of the book we will open up a further perspective into the diversity of actors and their mutual relations by several case studies on the northern European regional interface of the energy dialogue, including regional actors in Russia. The *regional politics of energy* is the third instance where we wish to open up some new perspectives (Table 1.1).

We suggest that each approach has its own premises with regard to what the main actors are, what they do and how, and why they do so. Whilst the approaches can in this manner be analytically separated, in individual research efforts we frequently derive from several approaches due to the numerous issue linkages and multifaceted nature of energy policy in general (Table 1.1). Before proceeding to what we will argue in the individual contributions by making use of these approaches, we will first probe in more detail into what we mean by our central concerns of 'European energy policy' and 'pan-European energy policy'.

7 E.g. Jack M. Hollander (ed.), *The Energy-Environment Connection* (Washington, DC: Island Press, 1992). Technical and technological energy research is not included here due to its indirect linkage to energy *policy*, which is here taken to refer to socially conditioned interaction.

Table 1.1 Approaches to energy policy

Approach	Premises			Example in this book
	Actors (who?)	**Agency (what?)**	**Interests (why?)**	
Energy diplomacy	States; regional blocs/institutions; political/business leaders	Summits; negotiations; agreements; treaties; conflict resolution	Patterns of dependence and interdependence; linking absolute economic gains and political stability	ch. 4 (Romanova) ch. 5 (Westphal) ch. 6 (Aalto and Tynkkynen)
Geopolitics and energy security	States; regional blocs/institutions; political leaders; state-bound companies	Energy and security policy; pipeline politics; energy shortages, cut-offs and crises	Exploitation of and protection against dependencies and vulnerabilities; relative economic gains and power accumulation	ch. 3 (Morozov) ch. 7 (Berg) ch. 8 (Tkachenko)
Energy economics and trade	States; regional blocs/institutions; companies; international financial institutions	Energy extraction, infrastructure, transport and trade; investment and finance	Rationality; absolute economic gains and benefit maximization	Most chapters to some extent
Energy and the environment	States; regional blocs/institutions; companies; political/business leaders; NGOs	Negotiations; agreements; treaties; expert meetings and reports; social mobilization	Ecological awareness and sustainability concerns; criticism of short-term economic rationality and traditional geopolitics	ch. 6 (Aalto and Tynkkynen) ch. 7 (Berg)
Political sociology of energy	States; regional blocs/institutions; companies; political/business leaders; NGOs	Political and social factors enabling and constraining energy extraction, transport, trade and exploitation	Intertwining of interests and identities, as well as social and material structures; linkage politics	ch. 2 (Aalto) ch. 3 (Morozov)
Bureaucratic politics of energy	States; regional blocs/institutions; companies; policy-makers and officials in state institutions	Bureaucratic struggle and clash of sectorally and functionally differentiated institutions	Sectoral rationality; rent-seeking; balance of intended and unintended consequences in interest formation	ch. 4 (Romanova) ch. 5 (Westphal) ch. 7 (Berg) ch. 8 (Tkachenko)
Regional politics of energy	States; regional institutions and administrations, interest groups and NGOs	Centre-region and interregional struggles in energy exploitation, transit and trade	Regional rationality; rent-seeking; federal and regional development	ch. 4 (Romanova) ch. 8 (Tkachenko)

Questions of European Energy Policy

This volume's aim of contributing to European energy debates necessarily raises the wider question of the extent to which there today is a European energy policy; or preferably, as we would rather frame it, the extent to which there is, or should be, a *pan*-European energy policy. Answers to these questions naturally depend on what is meant by the term *policy*. Deriving from James Anderson, Marvin Soroos defines policy as:

> … a purposeful course of action designed and implemented with the objective of shaping future outcomes in ways that will be more desirable than would otherwise be expected. Discrete actions, in and of themselves, are normally not considered policies. On the contrary, a policy establishes criteria for deciding on an entire series of actions.[8]

Another way of putting this would be to say that a policy requires a degree of direction, purpose and stability of politics in a given realm such as energy.[9] Of course, this definition of the term policy does not imply the presence of a monolithic policy shaper and implementing body. In practice what we call policies are often formed in murky conditions defined by bureaucratic struggles, mutually conflicting problem definitions and internal contradictions as suggested by the bureaucratic politics approach (Table 1.1).[10] Added to this is the environment where policies need to be created, often characterized by uncertainty, complexity and conflicting voices among stakeholders and other groups affected by the policy.[11]

There definitely is a relatively long-standing aim of building the required direction and purpose amounting to thus understood 'policy' that would provide for collectively good outcomes in European energy politics. The main source of this drive is the process of European integration, despite of the fact that the results attained in the energy sphere since the end of the Cold War have not been as impressive as many have wished, and as has been seen in many other spheres of European integration. But apart from the problems of internal EU integration and the level of EU regulation, the EU needs to pay attention to its relations with its most important energy suppliers, first and foremost Russia, followed by Norway, Algeria and Saudi Arabia (Table 1.2). Another issue the Union needs to keep on the table comprises important energy transit countries in the EU–Russian borderlands, most notably Belarus and Ukraine, whilst also paying attention to transit issues within its own territory.

---

8 Marvin S. Soroos, *Beyond Sovereignty: the Challenge of Global Policy* (Columbia, SC: University of South Carolina Press, 1989), p. 19.

9 Kari Palonen, 'Introduction: Reading the Political: From Policy and Polity to Politicking and Politicization', in Kari Palonen and Tuija Parvikko (eds), *Reading the Political: Exploring the Margins of Politics* (Helsinki: Finnish Political Science Association, 1993), p. 13.

10 E.g. Graham T. Allison, *Essence of Decision: Explaining the Cuban Missile Crisis* (Boston: Little, Brown and Company, 1971).

11 Emery Roe, *Narrative Policy Analysis: Theory and Practice* (Durham: Duke University Press, 1994), pp. 3–18.

In other words, when speaking of European energy policy one evidently encounters the issue of management of external energy supplies to the EU area. This is why we prefer to treat questions of European energy policy as only raising new questions of the possibility of a *pan-European energy policy*. This naturally is quite a different challenge from devising a national or *governmental policy* within a member state. It is also different from a *foreign policy* of a given member state, i.e., something that is directed beyond its own jurisdiction. Hence, what we are faced with is perhaps best understood as a *call for an energy policy within the wider European area.*[12]

Table 1.2 EU's external energy supplies

	Gas imports (2005, EU–25)	Oil imports (2005, EU–25)
Russia	40%*	36%**
Norway	25%	20%
Algeria	19%	–
Saudi Arabia	–	11%

* *Author estimate (37% without Estonia, Latvia, Lithuania and Slovenia; see Table 1.3).*
** *Including the former Soviet Union area where Russia is the main provider and transit country.*
Source: European Commission, 'Energy and Transport: Figures and Main Facts: Statistical Pocketbook 2006', <http://ec.europa.eu/dgs/energy_transport/figures/pocketbook/2006_en.htm> (accessed 16 February 2007).

The creation of a wider European policy of this kind requires support from governmental policies that in Europe still represent the main source of energy policy decision-making as will be explained below. Simultaneous support is needed from the foreign policies of individual EU member states and the Union's partners, most notably Russia, but also energy transit countries like Ukraine and Belarus. As implied, energy resources in the European continent are unequally distributed across states, making most EU members net importers of energy; in the case of gas, a very high dependence on Russian supplies is no rarity (Table 1.3). At the same time both in the east and west of Europe, many energy companies are state owned in part or in total. Therefore, the exploitation of energy resources between the consuming and producing countries often becomes an issue of foreign policy, and an instance of its intersection with other trade interests and interests of energy companies. But importantly, in the case of energy policy within the wider European area, all these policy elements should be conjoined, and accommodated with the somewhat painful emergence of a *common EU foreign policy* towards Russia that has been a frequent subject of critical debate.[13]

12 Though demanding, this is less than the *global policy* proposed in Soroos, *Beyond Sovereignty*, pp. 20–1.

13 See e.g. Hiski Haukkala, 'A Problematic "Strategic Partnership"', in Dov Lynch (ed.), *EU–Russia Security Dimensions* (Paris: European Union Institute for Security Studies, 2003);

We will next review in more detail the efforts during the past decade and a half of creating an internal EU energy policy. Then the discussion moves towards questions of externally oriented policies and the possibility of a pan-European energy policy where the EU–Russia energy dialogue for many reasons occupy a pivotal role, despite the pressures emanating from governmental policies and foreign policies of the EU, its members and Russia, not to mention other interest groups.

Table 1.3 Gas imports from Russia among some EU member states, 2004

EU member state	Share of total imports, %
Bulgaria	100
Estonia	100
Latvia	100
Lithuania	100
Finland	100
Romania	100
Slovakia	100
Greece	82
Austria	82
Hungary	81
Czech Republic	74
Poland	63
Slovenia	60
Germany	45
Italy	37
France	21

Source: European Commission, 'EU Energy Policy Data', SEC(2007) 12, Brussels; note that gas consuming EU member states producing gas themselves or importing it from elsewhere are not included. Neither is the share of gas in the energy mix of the listed states taken into account.

Step I: Principles of Internal EU Energy Policy

The origin of the efforts of creating an internal EU energy policy can be traced to the early stages of European integration (see Chapter 2). Yet, only in the aftermath of the 1970s oil crises were some lowest common denominator type guidelines initiated. Energy was initially left out of the single European market concept, and was only added there in 1988. Since then, the history of EU's energy policy represents a struggle conducted mainly by the EU Commission at the face of often heavy-

---

see also Debra Johnson and Paul Robinson (eds), *Perspectives on EU–Russia Relations* (London: Routledge, 2005).

handed involvement of governmental policies of member states.[14] However, three systematically pursued principles can be identified.

The first principle, *market rules and competitiveness*, reproduces the core idea and the strongest sector of EU integration, the single market. This is the policy sector where the Union is at its strongest in terms of its regulative capacity, and has also been named 'regulatory state', in order to differentiate it from how governmental policies of member states focus on distributive and redistributive functions alongside national level regulation.[15] And in conditions of high energy prices, market rules and regulation aimed at guaranteeing competition have assumed a new salience.

The EU Commission has gradually taken a more proactive line on implementing the market rules and competitiveness principle within the Union. Whilst competitive electricity and gas markets were established in EU directives in 1996 and 1999, respectively, and sharpened in the directive of 2003, in the majority of member states the implementation of these regulations remains slow and several bottlenecks prevail.[16] There are also strong protectionist pressures for maintaining national monopolies. The proposed takeovers of the Spanish Endesa by the German E.On and the French Suez by the Italian Enel in 2006 evoked national countermeasures. To amend the unsatisfactory state of the internal EU energy market, the EU Commission's 2006 Green Paper proposed the establishment of a European Energy Regulator. It also suggested a European energy grid code and further interconnection of European energy grids.[17] In the field of electricity, failures to meet the Union's investment targets for enhancing the transeuropean electricity networks have been frequent throughout the Union.[18] The new member states from Central and Eastern Europe (CEE) joined the European interconnected electric system by 2003, but in practice many technical handicaps and dependencies remain. For example, Estonia and Lithuania are for the time being mostly electricity self-sufficient, but yet,

---

14 Janne Haaland Matlary, 'The Nordic Countries and EU Membership: the Energy Factor', in Lee Miles (ed.), *The European Union and the Nordic Countries* (London: Routledge, 1996), pp. 239–41.

15 Giandomenico Majone, 'Regulatory Federalism in the European Community', *Environment and Planning C: Government and Policy*, 10 (1992): 299–316. For example the common agricultural policy and the structural funds represent redistributive EU policies where member state contributions are pooled. But the scale of such schemes is small compared to nation-states' typical redistributive functions through taxation. Members only pay around 1 per cent of their budget to the EU.

16 European Commission, 'The Commission to Act over EU Energy Markets', MEMO/06/481, 12 December 2006, <http://europa.eu/rapid/pressReleasesAction.do?reference=MEMO/06/481&format=HTML&aged=1&language=EN&guiLanguage=en> (accessed 23 March 2007); see also Table 9.3 in Chapter 9.

17 European Commission, 'Green Paper: A European Strategy for Sustainable, Competitive and Secure Energy'.

18 European Commission, 'Towards a Competitive and Regulated European Electricity and Gas Market', memo, <http://ec.europa.eu/energy/electricity/publications/doc/2004_07_09_memo_en.pdf> (accessed 24 October 2006), pp. 5–6.

alongside Latvia, they are technically and infrastructurally linked to the Soviet era electricity networks, and thus today, Russia.[19]

The second principle, *sustainable development*, is also very much characteristic of the EU but comes with a shorter history than the market. It was accepted as a common EU goal in the 1997 Amsterdam Treaty, although energy and environmental policy were linked within the EU already in 1990.[20] Environmental protection and the fight against climate change are policy fields in which EU member states have reached a stable consensus. This consensus suggests that energy saving and increased energy efficiency as well as a greater utilization of renewable energy resources will reduce greenhouse gas emissions by cutting the use of hydrocarbons, and at the same time save scarce resources. Consequently, in order to limit global warming to two degrees Centigrade above pre-industrial levels, the EU has undertaken several steps on the Union and member state levels to reduce carbon dioxide emissions by means of the EU emissions trading scheme, including the promotion of the Kyoto Protocol and its global scale mechanisms. In 2005, the Commission published a Green paper on energy efficiency that aims at increasing energy efficiency by 20 per cent. The Green Paper of 2006 combines sustainable development with innovation and new low carbon technologies.[21] The EU has become a worldwide promoter of linking environmental and climate issues with energy consumption and the use of new technologies.

The principle of *security of energy supplies* springs from the Union's worsening energy poverty (see also Chapter 2). Energy security measures within the EU have included crisis management mechanisms along the rules of the International Energy Agency (IEA), in which most EU member states belong to and where the EU participates as an observer. With regard to oil, the IEA rules and the single market legislation provide some guidelines for the maintenance of security stocks of crude oil and petroleum products on the national level as well as possibilities for anti-crisis measures. But the EU Commission is aware of how the mechanisms introduced in the early 1970s on the national level are no longer suited to the current EU oil market. In the Green Paper of 2000, the Commission tried to start a debate on the security of energy supplies in Europe, prompted by the expectations of increased energy imports dependency, growth of energy demand, the existing energy mix and decreased production within the EU. However, this Green Paper mainly focused on the demand side, mentioning also the possibility of energy savings.[22]

19 Diana Ürge-Vorsatz, Gergana Miladinova and Laszlo Paizs, 'Energy in Transition: From the Iron Curtain to the European Union', *Energy Policy*, 34 (2006), p. 2285. Estonia's use of oil shale may not be sustainable in the long run, and Lithuania's nuclear power plant in Ignalina is set to be closed down by 2009.

20 See Matlary, 'The Nordic Countries', p. 241.

21 European Commission, 'Green Paper: A European Strategy for Sustainable, Competitive and Secure Energy'; European Commission, 'Green Paper on Energy Efficiency: Doing More with Less', Brussels, 22 June 2005, COM(2005) 265 final; see also Kirsten Westphal, 'Energy Policy between Multilateral Governance and Geopolitics: Whither Europe?' *Internationale Politik und Gesellschaft*, 4 (2006): 44–62.

22 European Commission, 'Green Paper – Towards a European Strategy for the Security of Energy Supply', Brussels, November 2000, COM(2000) 769 final.

Already in 2002 the Commission complained about a lack of solidarity in particular in the use of strategic oil stocks. This was taken up in 2006 again, when the Commission suggested coordination and harmonization of the EU member states' oil and gas storage systems. Solidarity among the member states was defined in more general terms as decisive for improved energy security in the Green Paper of 2006 and in the energy package of 2007.[23]

* * * *

The principle of sustainable development implies an agenda that goes beyond not only governmental policy, but also EU policy, due to the transboundary nature of environmental threats. The security of supplies principle makes the external dimension even clearer due to the EU's energy poverty. But the EU has also tried to extend the market principle outside its own sphere of regulation. The first act where the insufficiency of internal EU integration was clearly recognized, was the case of the Energy Charter Treaty (ECT). This treaty is a logical starting point for examining in more detail the turning of the EU's energy policy into the external, and in particular, Russian direction.

Step II: The Energy Charter Treaty (ECT)

The three principles of internal EU energy policy were all in different forms present already in the ECT, which was formed on the basis of the non-binding 1991 European Energy Charter. After three years of negotiation the ECT was agreed on in December 1994, marking the first economic agreement intended to unite altogether 52 states ranging from the former Soviet Union, the former centrally planned CEE states, EU member states, plus Norway, Switzerland, Turkey, as well as Australia and Japan. China and Saudi Arabia are important observers of this relatively ambitious effort.

The ECT entered into force on 16 April 1998 as a binding treaty under public international law. It covers central areas of energy co-operation such as trade, investment, transit, dispute settlement, and energy efficiency. The treaty explicitly refers to the General Agreement on Trade and Tariffs (GATT) and incorporates in article 29 IIa the principle of equal treatment of national and foreign actors. In doing so, the treaty attempts to strengthen the market principle and multilateralism in the form of equal access to markets and networks. Ideally it aims at providing a basis for the creation of an open regional energy market under the principles of the GATT and its successor World Trade Organization (WTO).

The market principle in the wider European context is intended to provide a basis for contractual and trade relations in energy policy, i.e., to create a basis for the rule of law in this sphere. This facilitates smaller companies' participation in energy trade, as in general they are not capable of negotiating individual agreements with governments in the same way as European and Russian energy giants are. Yet, although the ECT was created as an international regime for liberalizing trade, investment and transit, it does not impose privatization in the energy sector or third

23 European Commission, 'An Energy Policy for Europe', p. 10.

party access to domestic pipeline networks on the *national* level. Moreover, the ECT reaffirms national *sovereignty* over energy resources. National governments retain the right to determine the territory to be exploited, the right to define depletion and reserves policies, as well as the level of taxation, and the right to participate in exploration and production.[24]

Sovereignty-related considerations have been on the way to the ECT's full application. The treaty's main aim of creating a pan-European framework were dealt a severe blow when the most important energy supplier in this regional space, Russia, refused to ratify the ECT due to its Protocol on Transit, which is one of the centre pieces of the treaty. The Russian policy fell together with the strategy of Gazprom, whose nearly full natural gas monopoly in Russia is strengthened by its position as the operator of the Russian gas pipeline network and many other pipelines within the former Soviet territory. By contrast, in the EU markets Gazprom acts as a multinational globalized actor. It has become involved in marketing and distribution of its gas instead of simply delivering it to the border of the customer country. In a well-balanced 'downstreaming' policy, it has established trading and transportation firms in eastern and western European countries.[25] At the same time the Gazprom-controlled Russian natural gas grid and its export infrastructure are pivotal to the energy exporters in the former Soviet Union area who are eyeing supplies to the EU direction.

The precise sovereignty related issue here is that the ECT also regulates transit issues (article 7) and thus the access to *export* pipeline networks under the principles of freedom of transit and non-discrimination. In cases of disputes the treaty foresees international arbitration. The treaty and its transit protocol would oblige Russia to implement the principles of freedom of transit without distinction of the origin, destination or ownership of energy, and non-discriminatory pricing. But owing to its control over the domestic and the export natural gas pipeline infrastructure, Gazprom, and with it, Russia, enjoy a very favourable position in the energy transit from the gas fields and reserves in the Caspian region that account for about 4–5 per cent of global reserves.[26] Thus it is clear that ratification of the ECT and its transit protocol would simply not be in Gazprom's or Russia's interests. The consequent sovereignty bound policy of Russia is aimed at maintaining its position as Europe's chief supplier.

Step III: The EU–Russia Energy Dialogue

When it became evident that the ECT would not work as expected in relation to Russia, the EU made a further attempt to secure its energy supplies by proposing the EU–Russia energy dialogue in 2000. The Union received a positive response

---

24 Energy Charter Secretary, 'The Energy Charter Treaty: A Reader's Guide' (2002) <www.encharter.org>, p. 7.

25 For more detail, see Kirsten Westphal, *Russische Energiepolitik* (Baden-Baden: Nomos, 2000), pp. 98–106.

26 However, note that Gazprom correspondingly needs to purchase energy transit in the European direction in Ukraine, Belarus, Poland, the Czech Republic and Slovakia.

from Russia, making this sphere the best advancing sector of EU–Russia relations that today are often discussed under the rubric of 'strategic partnership'.[27] Three objectives can be identified for the EU–Russia energy dialogue.

First, the main aim of the energy dialogue is the typical diplomatic one of facilitating the flows of energy trade and investment by providing a political and institutional framework for increasing EU–Russia energy trade. Both parties are motivated for this as a result of the strong interdependence created by the EU's need to secure its supplies from Russia, and Russia's need to secure its energy demand from the EU, where it sells around 60 per cent of its exports. The mutual motivation for co-ordination is enhanced by how the two markets are profoundly different.

In the EU side, as discussed above, the Commission is determined to create a real competitive market. It is not unreasonable to expect this aim to succeed, albeit slowly, even at the face of considerable pressures for sovereignty and national protection on the part of member states. At the same time, EU based companies strive for full access to the Russian energy market, where they can provide badly needed investments (see also Chapters 2, 3 and 5). The Russian market, by contrast, is closed in the gas sector. Any efforts of breaking Gazprom's domestic monopoly would have important implications on the Russian government's foreign trade policy, and in fact, would deprive it of a powerful foreign policy weapon. Although Gazprom is very active in joint ventures and gas trading, and has transmission companies in Europe, as mentioned above, it does not operate on the basis of reciprocity.[28] But in the oil sector, European and US companies have joint ventures with Russian firms, as have Chinese and Japanese companies outside the wider European area.[29] The overall tendency that emerged especially during the second term of president Putin in Russia is, however, not towards a greater opening but rather vice versa (see Chapters 3 and 8).

Second, apart from the market logic, as a wishful side-effect on the EU's side, the energy dialogue represents a means of adjusting the divergent positions of EU member states in the energy trade with Russia. The 2004 agreed on Constitutional Treaty of the EU would have given the Union a shared competence in energy questions with member states. But in its absence, despite the discussed three principles of energy policy that the EU promotes, decisions pertaining to this sphere are only made within the sectors of the environment, technology, competition, and internal markets, with EU-designed transport corridors and logistics chains also entering

27 Within this framework, the energy dialogue is the dominant aspect of the Common Economic Space (CES) between the EU and Russia, which was only conceived in 2003 alongside the other common spaces on internal security, external security, and on education, research and cultural issues; see Pami Aalto, 'The EU, Russia and the Problem of Community', in Hartmut Mayer and Henri Vogt (eds), *A Responsible Europe? Ethical Foundations of EU External Affairs* (Basingstoke: Palgrave, 2006).

28 Margarita M. Balmaceda, 'EU Energy Policy and Future European Energy Markets: Consequences for the Central and East European States', Arbeitspapiere, MZES 42/2002, p. 42.

29 For more detail, see Aleksanteri Institute Eurasia Energy Group, 'The Energy Dynamic on the Borders of the EU: Internal Dynamics of the Russian Energy Sector', policy memo, <www.helsinki.fi/aleksanteri/energy>.

the picture.[30] This implies inability to establish instruments for coordinating energy policies along the lines the EU has developed in other areas, such as common EU standards, intervention funds and decision-making mechanisms.[31]

Third, for the Russian party, the EU–Russia energy dialogue represents at least an opportunity of sorts for attracting investments in order to maintain and eventually expand its energy exports, modernize its decaying energy infrastructure,[32] and create better conditions for the country's economy in order to move away from a 'third world' model of merely exporting natural resources in conditions of high energy prices, and towards a high technology based model with energy proceeds used for building an advanced knowledge-based society.[33] But just like the EU side's secondary objective of finding a unified voice in its external energy relations vis-à-vis Russia remains unfulfilled, the realization of this goal in the Russian side is by no means straightforward. This assessment makes the case for examining the main advances and shortcomings on both sides in more detail.

Wider Questions of the EU and Russia

The wider political sociology of EU–Russia energy relations is explored in Chapters 2 and 3. Chapter 3 by Pami Aalto sheds light on the EU party by first looking at the past and present of European energy debates and how they tie into European integration. The Chapter then introduces narrative policy analysis in order to outline the main narratives commonly told by European policy makers and analysts to justify and legitimize the European integration project. The first narrative takes European integration primarily as an economic co-operation project motivated by mutual material interests by the participating states, including their interest in joint management of energy resources. The second narrative takes European integration primarily as a political or normative peace project motivated by mutual interests in avoiding future conflicts among the participating states. Representatives of EU institutions have so far chiefly used the economic co-operation narrative in order to refer to and justify the EU–Russia energy dialogue. The energy dialogue has remained dominated by economic and technical concerns despite the fact that crucial questions pertaining to the EU's future can be related to it: security of energy supplies, relations with Russia, and the question of Russia's place in Europe. The dialogue has thus remained safe from widespread politicization by the larger bulk of actors involved in and affected by European energy policy. However, the recent

30 See also Kirsten Westphal, 'Handlungsbedarf: Die Energiepolitik der Europäischen Union', *Osteuropa*, 9–10 (2004): 39–54.

31 Michel Chatelus, 'Oil Security and Relations between the Gulf Countries and the EU', in Michel Chatelus, Giacomo Luciani and Eberhard Rhein (eds), *EU-GCC Co-operation in the Field of Energy*, EUI Policy Paper 02/3, p. 7.

32 In 2000, 73 per cent of the oil transit network was over 20 years old and 41 per cent over 31 years old; for more detail, see Aleksanteri Institute Eurasia Energy Group, 'The Energy Dynamic on the Borders of the EU'.

33 Voloshin, 'EU–Russia Energy Dialogue', p. 28; see also Aleksanteri Institute Eurasia Energy Group, 'The Energy Dynamic on the Borders of the EU'.

problems the EU and Russia have encountered with important energy transport countries like Ukraine and Poland point towards a new era where member states become more keenly involved, and where large businesses become more wary of non-governmental organizations (NGOs) and other societal pressure groups developing an interest in energy questions.

Chapter 3 by Viatcheslav Morozov puts the energy dialogue into the wider context of competing visions for Europe. Both the EU and Russia face the need to make difficult decisions on the fundamental principles of their future development, and it seems that both are unable to do that, instead opting for keeping things as they are. Brussels all too often continues to use the logic of enlargement in its relations with Russia, including the principle of conditionality, amounting to what sometimes in scholarly literature is conceptualized as an imperial logic of absorption and assimilation. This is counterproductive, as Russia is unlikely to apply for membership any time soon. Consequently, its interests in the energy sphere do not follow blindly what the EU would like to see in terms of market opening and improving the investment climate. Disputes like these initially look technical, but eventually turn into fundamentally political questions of what the EU and Russia represent as political projects, and of whose models are being adopted in the institutionalization of their mutual relations.

Morozov suggests that this political impasse includes a fundamental political identity conflict between the EU and Russia. It is plunging Moscow into the nostalgia for the Soviet empire and prompting it to make decisions on the basis of traditional geopolitical logic. This leaves very little space for using the energy dialogue as one tool for Russia's industrial re-structuration, overall modernization and transformation. In this manner, opportunities within the EU–Russia energy dialogue remain unexploited, with both parties maintaining too much of a mutually exclusive, centralist way of thinking. This also hints that the regional dimension of the issue remains far too neglected, although some space for action remains there.

Questions of Levels

Chapter 4 by Tatiana Romanova shifts the discussion into energy diplomacy and bureaucratic politics. The chapter suggests that the strategic partnership level is the determining one within the dialogue, in setting its priorities and its politico-legal framework. Romanova revisits the EU Commission's efforts of making the Union speak with one voice and judges the dialogue's focus on technical issues representing a lowest common denominator between the Commission, member states and businesses. On the Russian side, the debate has become more government-led during Putin's era. The Russian party's growing assertiveness may alter its image of being largely a reactive player defending its interests in a short-term tactical game.

Figure 1.1 Major oil pipelines from Russia to Europe

Source: Leibniz-Institute for Regional Geography, Leipzig, 2007.
Copyright: Leibniz-Institute for Regional Geography, 2007.

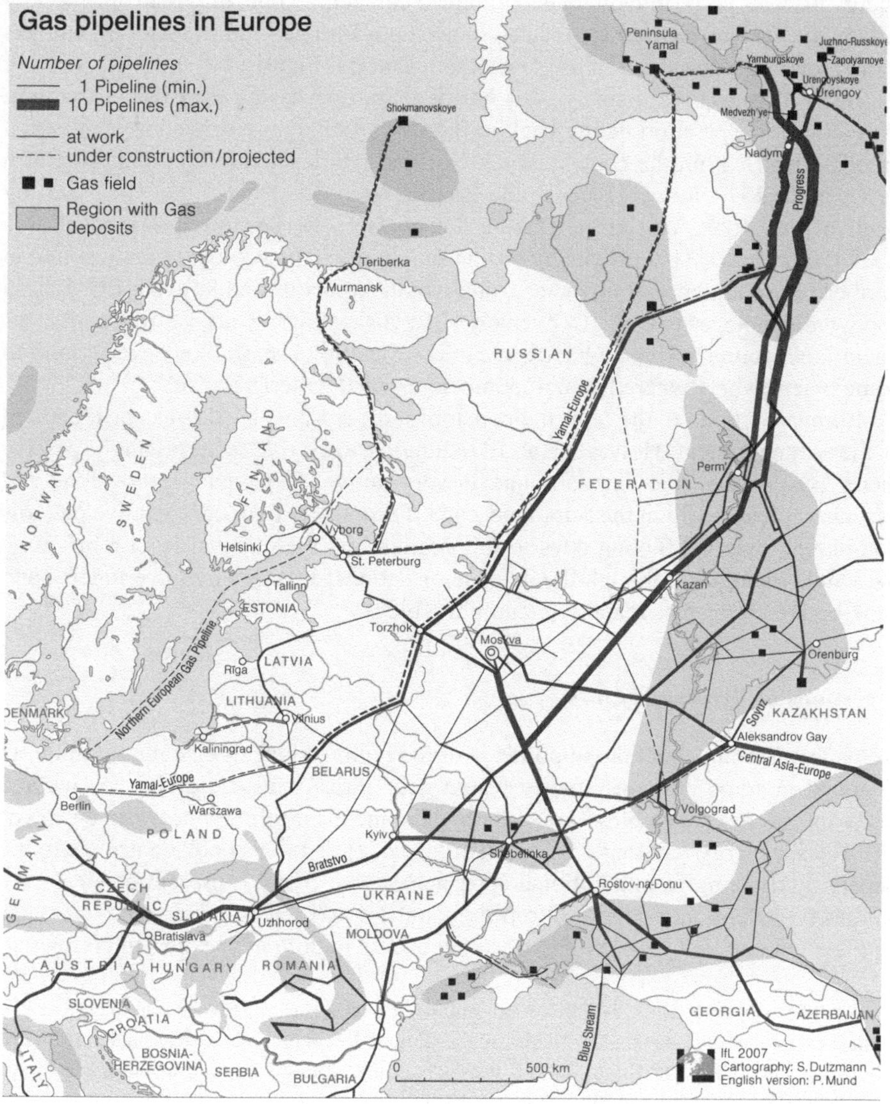

Figure 1.2 **Major gas pipelines from Russia to Europe**

Source: Leibniz-Institute for Regional Geography, Leipzig, 2007.
Copyright: Leibniz-Institute for Regional Geography, 2007.

Romanova's key point is to make the case for a regionalist approach by arguing that the strategic level decisions are actually put into practice at the regional level. She argues that a special emphasis on the regional level of northern Europe is particularly useful for several reasons: first, this is the only area where the EU and Russian energy markets come into direct geographical contact; second, important and largely untapped energy resources are located in northern Europe;[34] third, major oil and gas transit routes go through or close to northern Europe (Figures 1.1 and 1.2). A fourth reason is also implied, namely, that northern Europe has an established tradition of regional co-operation dating back to the early 1990s regionalization, and which is today dealt within the context of the Northern Dimension (ND) policy platform between the EU, Norway, Iceland and Russia. Here we arrive at the wider starting point of this book. We suggest taking the northern European regional level as a *crucial case study*:[35] if a promising setting for energy policy co-operation is found in northern Europe, there, if anywhere, one would expect to find how the EU–Russia energy dialogue can work. Conversely, if a disappointing degree of energy co-operation is found here, a higher degree of co-operation is most likely even harder to come by elsewhere where less co-operative regional patterns prevail.

Romanova finds all the same major actors at the regional level as can be identified at the strategic level. However, the EU Commission's role is surprisingly passive there. By contrast, the Russian state is very active. Regional organizations and various companies fill in the actor landscape. The chapter finds co-operative patterns dominating over conflicting ones in this region, and the regional level reinforcing the whole dialogue, although the defence of national interests by governments and state-bound companies remains a stumbling block.

Questions of Relevant Actors

Romanova's point about the importance of the regional level is also more implicitly assumed in some previous studies where geopolitical and energy economics and trade approaches are employed in order to examine the north European dimension of EU–Russia energy relations.[36] The remaining chapters to this book contribute to the debate by deriving from additional approaches. The chapters present detailed case studies on individual northern European countries and actors therein, and make use of expert literature, news sources, documents, and importantly, expert interviews.

---

34 These include Russia's Shtokman gas fields in the Barents Sea as well as Norway's resources in the high north; oil fields in the Timan Pechora area in Russia's northwest, extending into the Komi Republic and the Nenets autonomous area, and major gas fields north from Timan Pechora in the Yamal peninsula.

35 This term is further developed in Harry Eckstein, 'Case Study and Theory in Political Science', in Fred I. Greenstein and Nick W. Polsby (eds), *Handbook of Political Science, vol 7: Strategies of Enquiry* (Reading, MA: Addison–Wesley, 1975).

36 For energy economics and trade, see Tero Lausala, 'The Role of Energy in the Northern Dimension', in Lassi Heininen (ed.), *Northern Borders and Security – Dimensions for Regional Cooperation and Interdependence* (Turku: Turku School of Economics and Business Administration, Business Research and Development Centre, 2002); for geopolitical approaches see Keith C. Smith, *Russian Energy Politics in the Baltics, Poland and Ukraine: A New Stealth Imperialism?* (Washington, NY: Center for Strategic and International Studies, 2004); for approaches combining geopolitics and economics, see Juhani Laurila, 'Transit Transport between Russia and the European Union in Light of Geopolitics and Economics', *Emerging Markets Finance and Trade*, 39 (2003): 27–57.

Chapter 5 by Kirsten Westphal proceeds from an energy diplomacy perspective. The chapter introduces German perspectives into the EU–Russia energy dialogue that are relevant both for the strategic level and the regional level due to Germany's role in a key north European energy project, the Nord Stream pipeline project through the Baltic Sea. This project made German–Russian energy relations increasingly bilateral, even acquiring a personal character between the Russian president and the German Chancellor Schröder, coupled with asset swaps and close co-operation between Gazprom and German energy companies. This is in contradiction to Germany's reliance on multilateralization in EU energy policies throughout the 1990s, although it continues to conduct sustainability policies through the EU. These sovereignty bound German choices are found somewhat questionable even from the point of view of Germany's own energy security, let alone the concerns within the wider European area by Germany's neighbours and transit countries. In addition, the economic dimension and fairly particular business interests are promoted at the expense of political dialogue. Westphal calls Germany's post-Schröder era governments to return to the line within the EU, to help genuinely diversify German and EU supplies by means of infrastructure projects, and to work in order to integrate Russia into jointly agreed multilateral frameworks.

Chapter 6 by Pami Aalto and Nina Tynkkynen looks at the Nordic states by employing an energy diplomacy approach combined with a focus on energy and the environment. Denmark is found a somewhat detached actor in the energy dialogue largely thanks to its energy self-sufficiency. Sweden's energy political choices have made looking towards Russian supplies a lot more attractive. Norway is both a competitor and partner to Russian actors in being a major producer. Like Russia, it also has a strongly state-guided energy sector despite their very different institutional set-up and Norway's participation in the internal energy market of the EU. Norway's ability to involve the sustainability principle into its own, separate energy dialogue with Russia is proposed as potentially a significant contribution to the realization of the principle in north European energy politics, something that all Nordics support strongly. Finland is 40 per cent dependent on Russian supplies. Finnish governmental and NGO actors criticize the dialogue's quality at the strategic level and favour regional level co-operation and down-to-earth projects, especially for the taming of environmental threats resulting from energy production and transportation. Although Romanova's chapter deems northern Europe's regional level on the whole as contributing positively to the EU–Russia energy dialogue, Aalto and Tynkkynen find some resonance into this picture and the Nordic countries having mutually conflicting interests, with security politics also interfering in the debates.

The security political link is amplified by Eiki Berg in Chapter 7. The chapter focuses on the Baltic states' positions in east–west energy trade and diplomacy, and illustrates well their dual roles by looking at both geopolitical and environmental issues. The Baltic states are prime examples of the Russian strategies of acquiring assets in the EU area. However, due to the conflictual nature of overall relations between Russia and the Baltic states, not all Russian approaches in the Baltic markets are appreciated. A case in point is the saga of the Ventspils oil terminal in Latvia which lost its pipeline deliveries from Russia with the building of new ports into the Leningrad *oblast*. Simultaneously Estonian ports have managed to do fairly well even in the absence of any governmental policy on transit questions. In Lithuania, economic vulnerability with regard to energy deliveries from Russia is treated as a security issue that governmental policy should address. Overall, Berg concludes that each country relies on different infrastructural assets, political strategies vis-à-vis east–west energy trade, and represents a different set of bureaucratic and other domestic actors divided into transit lobbies, policy makers and environmentalists.

The final case study in Chapter 8 by Stanislav Tkachenko returns to Russia. The chapter makes an important contribution by providing an insightful and nuanced analysis of the bureaucratic and regional politics on the Russian side. Tkachenko takes issue with the now conventional geopolitical wisdom of the Russian energy sector becoming more state-governed under Putin. Healthy comparisons should be made to other energy producing countries where state involvement is much more dominant. Gazprom's almost total gas monopoly should be dissociated from the genuine competition in the oil sector, where foreign companies continue to be involved regardless of their shrinking action space. Tkachenko sheds new light on the notions of Putin's Russia as one energy superpower in the making, by pointing at conflicts of interests among Russian ministries, the centre and the regions, and at the competition between companies. Finally, although Tkachenko acknowledges that centralization remains a lasting trend in Russia's energy sector, he hints that this may be a better situation than the rent-seeking manoeuvring by bureaucratic actors in the energy sector that was characteristic of the Yeltsin era. But in the interests of good economic governance, Tkachenko calls for a balance between the public and private sector.

Chapter 9 draws the threads together and summarizes the main findings of the book with regard to what actor landscapes and interest formation patterns the individual chapters help to bring up with their varying combinations of theoretical approaches. The in-depth case studies of the realization of the energy dialogue on northern Europe's regional level suggest four sets of main problems obstructing the formation of a pan-European energy policy. These tie into the issues of sovereignty, asymmetry, bilateralism and imperfect commitment to principles within the EU and beyond. The spreading of the principles of EU energy policy into the wider European area towards Russia, but partly also Norway, is found problematic. Two policy options for the EU–Russia energy dialogue's further development are outlined. The first one keeps the dialogue in its present confines. The second uses a political sociology of energy policy approach to widen the dialogue's actor landscape and covered policy sectors in order to more openly acknowledge the spill-overs that we have found materializing already now. This second option poses fundamental questions to both parties but also opens up the energy dialogue's opportunities more fruitfully.

On the whole, we suggest in this book that energy policy is a far too important topic to be left either for policy makers, energy specialists, or to the turf wars of the academic community. Some of the contributions in this book prefer to stay within the theoretical confines of the bulk of existing energy policy research on EU–Russia relations; yet, they develop policy advocative edges as part of their analyses. Some others of us wish to recast important parts of the whole debate by deriving inspiration from new theoretical approaches to energy policy research. In this way, we wish to encourage *dialogue* both at the policy and theoretical levels and ensure different voices are heard.

The dialogue we suggest should include the already existing perspectives: energy diplomacy, energy economics and trade, geopolitical approaches and environmental research. In addition, we introduce political sociology approaches, as well as bureaucratic and regional politics approaches into energy policy analysis in order to open up the wider linkages of energy policy that help to sustain political and social life. In this way, the widened research agenda in this book becomes loosely linked to what some in the field of international relations have in recent years started to refer to as 'pragmatist approaches', connoting methodological pluralism, disciplinary tolerance, and importantly, dialogism, as a loudly pronounced foundation for a multi-perspectival enquiry.[37] In this spirit, we wish to leave the field open for competing perspectives, and trust that our work invites further dialogue.

37 See e.g. Editors of Millennium, 'Editorial Note', *Millennium: Journal of International Relations*, 31/3 (2002): iii–iv; also Helena Rytövuori-Apunen, 'Forget 'Post-Positivist' IR!: The Legacy of IR Theory as the Locus for a Pragmatist Turn', *Cooperation and Conflict*, 40/2 (2005): 147–77.

Chapter 2

The EU–Russia Energy Dialogue and the Future of European Integration: From Economic to Politico - Normative Narratives

Pami Aalto

Introduction

This chapter discusses the wider ramifications of the EU–Russia energy dialogue for the present and future of European integration. The chapter first briefly looks into European energy debates in the early 1980s and then makes a comparison to the situation in the new millennium. During these two decades, the EU emerged as a key participant in European energy debates and became an increasingly salient reference point for policy-making, simultaneously as Russia's energy supplies gradually became pivotal. Against this backdrop of the EU and Russia having become key parameters of European energy policy, it is conceptualized what elements of the past still remain and what is new. The conceptualization leads to my main theoretical suggestion that in order to navigate through present-day European energy debates, policy makers and analysts in the EU area need to develop an idea of the wider context in which energy issues are embedded, including an idea of what European integration stands for, and what is the present and future place of energy policy and the EU–Russia energy dialogue within it. In other words, they need to develop workable narratives of European integration to guide their thinking on how to gear the EU–Russia energy dialogue and to make it intelligible.

Putting forth this narrative approach represents an effort of probing into the wider political sociology of European energy policy. In addition to the EU Commission and representatives of the Russian federation, and their policies, this approach also highlights interested and affected groups in the EU side, especially with regard to what sort of an actor landscape may be ahead in European energy policy.

European Energy Debates in the 1980s

In 1981, at a time when the EU had not yet emerged as an actor in European energy debates and when only a few ingredients for a European-wide energy policy existed, a volume ominously entitled *The European Transition from Oil* was published. In the introductory text to the volume Gerald Foley and Måns Lönnroth noted that the decades following the Second World War witnessed Western Europe turning to oil in order to keep the wheels of post-war reconstruction running. Sustaining the much needed growth of the economy and trade in war-torn Europe presupposed a concomitant increase in energy supply. This was possible by turning to oil as was done in the US, and thus lessening the reliance on coal that until the early 1950s had supplied 90 per cent of Europe's energy needs. Oil was technologically relatively easy to introduce. Exploration, production and commercial exploitation was at the hands of competent, mostly US based companies controlling the chains of operations from oil drilling to retail distribution. Most of the oil that first started to flow into Europe came from just liberated colonies with few opportunities to object to how their resources were exploited. Oil was also a unique energy source in the way it stimulated new industries like the automobile and motor manufacturing.[1]

Foley and Lönnroth put forth the challenge for Western Europe to start the long-term process of transition *from* oil into other energy resources. Such a perceived need after the wide-scale adoption of oil during the decades of the Cold War was conditioned by what had started to look depleting oil resources. The rate at which new oil fields were found at the turn of the 1970s and 1980s did not seem adequate to respond to any significant increases in consumption. Oil producers' willingness to produce more and with buyer-friendly prices was also questionable after the oil crises and emergence of a more assertive Organization for Petroleum Exporting Countries (OPEC) in the 1970s. Moreover, it was not at all certain how much oil Europe would be able to obtain in competition with the US, Japan and the rest of the world.[2]

Returning to coal did not seem a particularly attractive option either, as investing into new ports, railways, trains, storage facilities and power stations would require major funds and political support exactly when these commodities were becoming much more difficult to obtain due to rising environmental concerns about coal-based energy production. Importing natural gas on a large scale from Western Europe's then antithesis, the Soviet Union, or North Africa, neither looked a risk-free option at the face of indigenous west European production falling sharply in the Netherlands, which was then the world's largest natural gas exporter. Hydropower had its evident limitations of scale as it has today. Nuclear power's significance was almost negligible in the overall European context. Rising oil prices were bound to hamper economic growth, incur capital shortages, inflation, decreased demand for electricity and a poor investment climate for this capital intensive mode of energy production. Finally, new renewable sources of energy were acknowledged to be decades away from large scale commercially viable exploitation.[3]

These limitations in available and foreseeable energy sources were aggravated by the poor outlook for long-term economic growth in Europe. And it is precisely

1 Gerard Foley and Måns Lönnroth, 'The European Transition from Oil: Mapping the Landscape', in Gordon T. Goodman, Lars A. Kristoferson and Jack M. Hollander (eds), *The European Transition from Oil: Societal Impacts and Constraints on Energy Policy* (London: Academic Press, 1981), pp. 3–4; 14–15.

2 Foley and Lönnroth, 'The European Transition from Oil', pp. 7–9.

3 Ibid., pp. 9–14.

good growth that is needed for seriously re-orienting energy policy. But at the same time, the dilemma for energy policy decision-makers is that higher growth often brings with it increased energy consumption, since the two tend to go hand in hand.[4] Importantly, the context, substance and problems of European energy debates are not entirely dissimilar two and half decades after the publication of Foley and Lönnroth's perceptive article, although there now is one more participant in the game, the EU and its integration machine the European Commission (see Chapter 1).

European Energy Debates in the New Millennium

The mid-term and even long-term economic outlook is not necessarily much more promising for Europe than it was in the early 1980s. Western Europe's growth in the first six years of the new millennium was sluggish. Its major economies, which are also motors of EU economy, delayed badly needed structural reforms and ignored the fiscal policy stipulations of the European Monetary Union (EMU).[5] France and Germany in particular were criticized for their inflexible labour markets and for their budget deficits. Italy had problems of cutting its budget deficit under the 3 per cent mark of its Gross Domestic Product (GDP), a goal ensuing from the EMU, and was lacking drive in its economy. Despite better growth in the Nordic members and the new CEE member states, this means that the whole EU, the entity built on the foundations of the west European integration project, is evidently failing to reach the target set out in the EU's 2000 agreed-on Lisbon strategy. Instead of becoming the most competitive and dynamic knowledge-driven economy in the world by 2010, the Union is in danger of becoming bypassed by more vibrant economic areas in the Americas and Asia. What in relative terms is a poor economic outlook suggests limits for any notable re-orientation of European energy policy, where many of the old issues simply refuse to die away.

Concerns about the long-term future of oil-based economy continue to exist as they did in the 1980s. But the time horizon for their actualization is extended somewhat further into the new millennium, with the precise date depending on the particular premises and perspectives adopted in each prognosis. According to most alarming estimates by the Association for the Study of Peak Oil and Gas (ASPO), around 2010, global oil production is set to peak, with the EU area among others then staring at the mid-point of resource depletion.[6] The intergovernmental organization IEA, for its part, argues that the earth's presently exploitable energy resources, most of which are carbon dioxides like oil, will suffice to meet demand until 2030, although

4 Ibid., p. 4; 15.

5 E.g. Paola Subacchi, 'Reforming Economic Governance in Europe: Exploring the Road to Effective Coordination', *International Affairs*, 81/4 (2005): 741–55.

6 See Susanne Peters, 'Courting Future Resource Conflict: The Shortcomings of Western Response Strategies to New Energy Vulnerabilities', *Energy Exploration & Exploitation*, 21/1 (2003): 32–3; see also <www.peakoil.net>.

it is less certain at what cost.[7] But it must be kept in mind that any prediction on oil production is problematic unless we specify the underlying assumptions.[8]

What is certain is that compared to the 1980s, we are not any more as pronouncedly speaking of oil provision from the Middle East and OPEC. The Soviet Union's successor state, Russia, has emerged as the EU area's single most important oil provider, accounting for about one fourth of the oil imports of the EU–25. Close to a half of the EU's natural gas imports come from Russia through pipelines built in the latter stages of the Cold War and thereafter despite the concerns on 'Soviet gas' anticipated by Foley and Lönnroth (see also Chapter 1). One third of the uranium used for the generation of nuclear energy in the EU area is also of Russian origin, and for example Finland buys electricity from Russia, with some other countries like Sweden weighing up electricity imports (see also Chapter 6).[9]

Returning to indigenous energy resources like coal is even less an option as it was in the early 1980s. However, coal continues to account for 30 per cent of electricity production within the EU. Due to strengthened environmental concerns related to coal use that are now linked to the greenhouse effect and global warming, it looks like maintaining and/or increasing the share of coal in Europe's energy mix would require improving the efficiency of the current, mostly over 20 years old coal-fired power plants and investing into the development of new, near-zero emission coal technologies that do not yet exist. But for the time being, in the absence of new technologies, there must be concerns of the long-term profitability and viability of coal-based energy production in the EU. These concerns are aggravated by the steadily increasing dependence on imported coal from around one third to 51 per cent by 2020.[10] As for nuclear energy, in 2006 Finland was practically the only EU country investing more into it, although many others were starting to consider the same, as did Russia, by announcing that construction work for ten new reactors will have started by 2015. The EU Commission remained fairly low key on nuclear energy in its 2007 document 'An Energy Policy for Europe'. The document outlined nuclear energy's benefits for a low carbon economy, energy security and mix, but simultaneously noted the contentious nuclear waste and decommissioning issues,

---

7 International Energy Agency, 'World Energy Outlook 2004', executive summary (2004), <http://www.iea.org/Textbase/npsum/WEO2004SUM.pdf>, p. 1.

8 These include what is meant by 'oil' (does it include unconventional oils?); reserves (proven reserves or remaining recoverable reserves, including probable and possible reserves); and the operational, technical and economic conditions under which oil can be produced; see Robert Mabro, 'The Peak Oil Theory', *Oxford Energy Comment*, September 2006.

9 The figures for the EU's oil and gas dependence vary according to source. For the cited EU's own estimate for 2000, see European Commission, 'Communication from the Commission to the Council and the European Parliament: The Energy Dialogue between the European Union and the Russian Federation between 2000 and 2004', Brussels, 13 December 2004, COM(2004) 777 final, <http://www.europa.eu.int/comm/energy/russia/reference_texts/doc/2004_0777_en.pdf> (accessed 28 December 2005).

10 International Energy Agency, 'Reducing Greenhouse Gar Emissions: The Potential of Coal' (2005), <http://www.iea.org/textbase/nppdf/free/2005/ciab.pdf>, pp. 11, 23–4.

and reserved nuclear energy decisions to member states.[11] Finally, renewable energy sources, mostly consisting of hydropower, have stayed a relatively small scale resource, accounting for less than one tenth of the consumption in the EU–25 partly as a result of low investment and partly due to limitations of geography.[12]

Although the concerns for energy supply are now fairly similar to the ones in the early 1980s, we have seen some new environmental and technological issues emerging more forcefully into the debates, as well as Russia's emergence as a major energy provider for the EU area. But there are also warnings of the risks associated with 'putting all eggs in the Russian basket' by means of the EU–Russia energy dialogue. In Susanne Peters' interpretation, the energy dialogue mainly pertains to the security of supplies principle of EU energy policy (see Chapter 1), and on that plane represents an effort to *reduce the risks associated with the EU's dependence on Russian energy resources*. The EU Commission acknowledged the EU area's dependence on external sources in its 2000 Green paper, without envisaging any efforts to increase its energy self-sufficiency or decrease the dependence. This acknowledgement was made regardless of serious doubts as to whether Russia will be capable of delivering enough and reliably in view of its limited reserves, high domestic demand and moderate production capacity of its energy companies especially in the mid-term.[13]

Estimates of Russia's share of the world's oil reserves vary from around 5 to slightly over 10 per cent. For gas, the same figure is put to almost one third. However, by 2015, a significant decrease in Russia's gas production may be in sight, as the maturing west Siberian fields start drying up. Exports through new pipelines to Japan, China and South Korea are also negotiated,[14] and doubts remain of the feasibility and cost of exploiting new fields in difficult climactic conditions in the Russian north and eastern Siberia (see also Chapters 6 and 9). Russia's continued, oil-fuelled growth figures since the beginning of the new millennium put question marks on sustainability, particularly regarding the extent to which energy savings as envisaged in Russia's energy strategy until 2020 will be possible. Another question is how to fend off the rising domestic demand in Russia that usually comes along with higher economic growth.[15]

The production capacity of Russian companies – in conditions of drying fields, decaying infrastructure and the somewhat risky taking into use of new, not as easily exploitable fields – could be heightened by means of foreign investment. Russia's energy strategy foresees a need for some 460–810 billion US dollars of foreign

11 European Commission, 'Communication from the Commission to the European Council and the European Parliament: An Energy Policy for Europe', 10 January 2007, Brussels, COM(2007) 1 final, pp. 17–18.

12 Presentation of Christian Cleutinx, European Commission coordinator for EU–Russian energy dialogue, Luxembourg, 25 October 2005, <http://www.europa.eu.int/comm/energy/russia/presentations/doc/2005_luxembourg_en.pdf> (accessed 29 December 2005).

13 Peters, 'Courting Future Resource Conflict', p. 36.

14 Aleksanteri Institute Eurasia Energy Group, 'The Energy Dynamic on the Borders of the EU: Internal Dynamics of the Russian Energy Sector', policy memo, <www.helsinki.fi/aleksanteri/energy>.

15 Peters, 'Courting Future Resource Conflict', pp. 37–9.

investment during 2001–20 (!), in order to ensure the country's long-term economic growth and to improve the natural resources infrastructure.[16] While this is a huge figure, so far, there are somewhat mixed signs of whether or not issues like the heavy-handed acquisition of the oil giant Yukos during 2003–4 by the Russian government and its Rosneft company deters away much-needed foreign investors. The same concerns the events in Russia's Far East during 2006–7, where the Russian authorities pressed for controlling stakes to be given to Gazprom in projects formerly led by foreign companies (see Chapter 3). In a word, with such concerns on Russian supplies, and with the Russo–Ukrainian gas transit dispute in 2005–6 (see below), the EU made supply diversification, alternative energy sources and energy efficiency issues more integral parts of its energy policy thinking. In 2007, the Commission set the target of making renewables to account for 20 per cent of EU energy mix.[17]

Conceptualizing European Energy Policy Debates

The short review so far shows that although a new political agent, the EU, has entered the field, and has started a process called energy dialogue with its declared strategic partner Russia mainly to secure its supplies, many of the old concerns of European energy policy remain more or less the same as they did back in 1981. These concerns can be conceptualized as comprising uncertainty, complexity and polarization.[18]

First, we are still speaking of *uncertainty* in relation to future energy supplies: the demand and supply sides still largely develop independently from each other despite their political representatives being in an ongoing 'dialogue'. And this uncertainty is made only greater by the unknowables related to the future development of European sustainability policies and pressures from the publics and environmental movements, and the development of EU–Russian relations. Whereas the sustainability issue most likely will strengthen in the future, the question regarding the energy dialogue is whether it lives a life of its own, or whether developments elsewhere in the EU–Russia partnership and its 'common spaces' project, including the Common Economic Space of which the dialogue is part of, override the 'simple energy relationship'.

Second, we are still speaking of *complexity*: as so well pointed out by Foley and Lönnroth, for energy policy makers, the numbers emerging from energy forecasts are not as important as are the assumptions that originally went into the calculations. This leads to the argument that comes across throughout this volume: security of energy supplies and the promotion of sustainability – which has become more and more the game of the day with continued resource scarcity and amplified environmental concerns about energy production and consumption – cannot be ensured only by relying on the market to handle things. Security of energy supplies, sustainability and

16 Ministry of Energy of the Russian Federation, 'The Summary of the Energy Strategy of Russia for the Period of Up to 2020', < http://ec.europa.eu/energy/russia/events/doc/2003_strategy_2020_en.pdf> (accessed 6 November 2006).

17 European Commission, 'An Energy Policy for Europe', p. 14.

18 This conceptualization is strongly inspired by Emery Roe's narrative policy analysis approach to difficult policy problems; see Emery Roe, *Narrative Policy Analysis: Theory and Practice* (Durham: Duke University Press, 1994), pp. 1–19.

market rules in energy policy *have to be seen in a wider context where energy forms just one part of the total picture*. Amid new technological solutions, and mobilization of resources, capital, skills and administrative capacity for their utilization, it is political support that continues to be 'at the heart of the machine'.[19]

Third, when in the early 1980s European energy policy did not exist as a politically realized idea, today its realization suffers from the increasing *polarization* within the EU. There are several lines of divisions after the eastern enlargement of 2004 and the French and Dutch publics' refusal to accept the Constitutional Treaty for the Union in referendums in 2005. These divisions extend from federalist-leaning 'Europe-maker' member states and groups, to doubters of the Constitutional Treaty, and to more general eurosceptics like the UK and Denmark; from governments of large EU member states like Germany looking to re-establish their lead, but many of whom remained paralyzed by their poor domestic popularity rates in 2005–7,[20] to small ones fearing to be overrun by the formation of 'centres of gravity' of EU integration; from northern to southern and eastern, each with their own special expectations; and from more to less energy import dependent members, with the new member states in many cases highly dependent on gas supplies from Russia (see Chapter 1).

Towards a Narrative Approach

What European policy-makers and analysts can do in the field of energy policy in the current – and in many senses also persisting – conditions of uncertainty, complexity and polarization? In short, in order to devise grounds for energy policy making in this sort of difficult conditions, and to underwrite and stabilize the assumptions that are necessary for a workable energy policy, I suggest that they will need to resort to *story-telling*. This is a very common form of political activity when scientifically and technically secure information is not available, as it most often is not, and when policies nevertheless need to be developed, decisions made and funds assigned. Therefore in the remainder of this chapter I will attempt to locate the *wider narrative context* from which European policy makers and analysts need to derive in order to situate the EU–Russia energy dialogue as part of their day-to day professional activities and make it something that the EU should be engaged in. This means developing an idea of what European integration is all about, including its past, present and future, and then fitting the energy dialogue in.[21] By means of this narrative approach, the EU–Russia energy dialogue becomes understood as part of a larger story in the writing, and where two parties are trying to address mutual

19 Foley and Lönnroth, 'The European Transition from Oil', pp. 6, 16, 19–21.

20 Germany got a new government to replace Gerhard Schröder's weakened and long-serving cabinet in late 2005 (see Chapter 5); Italy's PM Silvio Berlusconi gave way to former EU Commission President Romano Prodi in spring 2006, whilst new presidential elections were due in France after the two terms of Jacques Chirac in 2007. In 2006, Britain's Tony Blair announced his intention to step down as PM after a brief transition period.

21 For the approach of narrative policy analysis loosely followed here, see Emery Roe, *Narrative Policy Analysis: Theory and Practice* (Durham: Duke University Press, 1994), pp. 1–19.

concerns by means of a dialogue, whilst other interested actors and affected groups are joining in.

How do European policy-makers and analysts then make use of narratives in order to justify the EU–Russia energy dialogue and stabilize its underlying assumptions? Sometimes they make coherent use of only one narrative in order to convince their European-wide and national audiences and discredit a competing policy position or choice in issues of energy policy. Sometimes they combine different narratives and refer to them simultaneously in order to develop a more multifaceted and sophisticated story lending credence to a particular policy position or choice. In other words, different agents can use different narratives even in one and the same instance so as to strengthen the weight of their message. For this reason, in the present context narratives are best understood as *analytically separable* although they often are very much intermingled with each other in the political praxis of European integration.[22]

The analytical use of narratives represents a new political sociology approach to energy policy research, whereby texts and social interaction are assigned a pivotal role in policy formation. It is assumed that the language we use for its own part helps to construct our realities within the material constraints set by issues like availability of energy and other resources to sustain our living environment as well as political and everyday life. This is why diplomats attach so much attention to the words they use, and the importance of textual justification and contextualisation is no different for those responsible for running European energy policy. On the whole, it is granted that whilst stories structure the way we approach European energy debates, and lead us to perceive different prospects for EU energy policy, and by extension, pan-European energy policy, they are conditioned by material factors of energy availability, geography and international political economy. This makes some of the analytically possible options more likely than some others.

There are two main narratives of European integration that policy-makers and analysts can utilize in questions and debates pertaining to the energy dialogue. These narratives can be termed the *economic co-operation narrative* and the *politico-normative peace project narrative*. Whilst a book-length story could be told of each, it will suffice here to simply summarize their main assertions and refer to their support bases before moving into an analysis of their use, impact, and what sort of responses they can evoke.

The Economic Co-operation Narrative

This narrative is grounded in rationalistic premises of international political economy. The main actors within the narrative are EU member states, large companies based in the Union, especially state bound ones, as well as political and business leaders. The narrative locates the beginnings of European integration in the interest of the original six members to co-operate economically in the aftermath of the destructive Second World War. European level co-operation within the economic sector emerged

22 Cf. Iain McLean, 'Two Analytical Narratives about the History of the EU', *European Union Politics*, 4/4 (2003): 499–506.

as an attractive option for a collective response to a whole set of problems: very much valid perceptions of Europe's diminished role in the world economy, the dominating financial and economic strength of the US, fears of the socialist mode of production adopted within the Soviet Union and its east European satellites, as well as perceptions of a rapidly revived German economy after the demise of the abhorred Nazi regime, and concomitant severe domestic economic problems in other core European states.[23]

At the political plane, the German question looked a burning one in the eyes of European policy makers, many of whom had fresh memories of the war and of the then efforts of the allied to contain German power. At the background also lurked questions of the worsening conflict between the east and west in Europe.[24] The Schuman declaration of 1950, the resulting establishment of the European Coal and Steel Community (ECSC) in 1951, the 1957 Treaty of Rome creating the European Economic Community (EEC), and the related European Atomic Energy Community treaty (EURATOM) were deemed a suitable initial response to the challenge. The resulting institutional arrangements laid the basis for co-operation in the economic sector and energy production. In this reading, joint interest in energy issues formed a cornerstone of European integration in its beginnings. But at that stage, there was no energy crisis, nor were energy issues treated in security of supplies terms as they were in the early 1980 and are today.

In the 1960s, with Germany's *Einbindung* to Europe firmly put on track, the Gaullist regime in the other motor of European integration, France, and its state-centric ideology, plus its intergovernmentalist idea of European integration functioned as another factor which kept European integration on its economic track and prevented the practical conduct of integration from spilling too much into the political sector.[25] Since then, later entrants into the EU such as the UK and Denmark in the 1970s, and to a slightly lesser extent, Sweden and Austria in 1995, have expressed and kept alive this idea of European integration as connoting a vast free trade area with relatively little spill-overs to other sectors. These countries have even been prepared to become net contributors to the EU budget in order to enjoy the benefits the Union brings especially to the exporting sectors of their economies.[26] For large EU-based corporations and their leaders this story also represents the main tool by which to relate themselves in public to European integration.

At the more academic plane of story-telling, Andrew Moravcsik's liberal-intergovernmentalist approach to EU integration explains the EU's history in this fashion, primarily in terms of French, British and German economic interests and interest-formation, as they negotiate and bargain in grand Inter-Governmental

---

23 Derek Urwin, *The Community of Europe: A History of European Integration since 1945* (London: Longman, 1995), p. 12.

24 Desmond Dinan, *Europe Recast: A History of the European Union* (Basingstoke: Palgrave, 2004), p. 5; John Gillingham, *Coal, Steel and the Rebirth of Europe, 1945–1955* (Cambridge: Cambridge University Press, 1991).

25 Dinan, *Europe Recast*, p. 6.

26 Håkan Wiberg 'Adaptive Patterns and Their Deep Roots: a European Overview', in Hans Mouritzen, Ole Wæver and Håkan Wiberg (eds), *European Integration and National Adaptations: A Theoretical Inquiry* (Commack: Nova Science Publishers, 1996), pp. 46–8.

Conferences of the EU.[27] John Gillingham's recent work portrays the market principle as 'gradually supplant[ing] that of the state in Europe's long march from the economy of war to that of peace'.[28] In another influential variant of this type of narrative, Alan Milward asserts that '[I]t is the post-War economic and social forces which have shaped the European state which therefore need to be analyzed if the origins and purpose of the [European] Community are to be explained and questions about the past, present and future of European integration are to be answered'.[29] Finally, in their own time influential functionalist accounts of European integration of course famously implanted the thesis that economic co-operation – which evidently at the level of EU regulation started the whole integration endeavour – is prone to subsequently spilling over into new policy sectors through technical and incremental processes. This core claim of functionalism logically leads us to the politico-normative narratives.

The Politico-Normative Peace Project Narrative

The main actors in this narrative are states and political leaders. Elements of this narrative can also be located in the founding treaties of European integration. The bulk of the text in the founding treaties is relatively technical in its provisions for the practical integration in the sphere of atomic energy (EURATOM) and various aspects of overall economic integration (Treaty of Rome). But the preambles of these treaties contain weighty references to the term 'peace'. The Rome treaty even contains the phrase that turned out to be particularly influential in the subsequent argumentation on promoting further integration: the portrayal of the EEC as a means to 'an ever closer union among the peoples of Europe'.[30]

The roots of the peace project narrative can easily be located even much further in European history. The 'civilizing' effects of the Roman Empire represent one of the historical role models. From more recent times, an obvious pick is the function of European integration in taming the long-running Franco–German disagreements. As put by the EU in its own Europa website:

> For centuries, Europe was the scene of frequent and bloody wars. In the period 1870 to 1945, France and Germany fought each other three times, with terrible loss of life. A number of European leaders became convinced that the only way to secure a lasting peace between their countries was to unite them economically and politically.[31]

---

27 Andrew Moravcsik, *The Choice for Europe: Social Purpose and State Power from Messina to Maastricht* (Ithaca, NY: Cornell University Press, 1998).

28 John Gillingham, *European Integration, 1950–2003: Superstate or New Market Economy?* (Cambridge: Cambridge University Press, 2003), p. xii.

29 Alan Milward, with the assistance of G. Brennan and F. Romero, *The European Rescue of the Nation-State* (London: Routledge, 2003), p. xi.

30 See e.g. Dinan, *Europe Recast*, pp. 76–7.

31 European Union, 'The History of the European Union', <http://www.europa.eu.int/abc/history/index_en.htm> (accessed 30 December 2005).

This narrative of European integration as a politico-normative peace project motivated by mutual interests in avoiding future conflicts and wars among the participating states, is often used in popular contemporary political discourse. It is also strongly associated with the ideology of Europe-building. The political leaders of the original six member states Germany, France, Italy and the Benelux countries often refer to this story of European integration leading to a Deutschean security community where dependable expectations of peaceful resolution of conflicts prevail widely, and where resorting to force is therefore unthinkable as a policy option. Many political leaders of the new member states that joined in the 'big bang' enlargement of 2004 learned using elements of this narrative to their own advantage when the progress of the enlargement round appeared stalled.[32] For many NGO and civil society actors this story represents a pivotal justification for EU integration.

The scholarly variant of the story by Walter Lipgens suggests that the ideas of European federalist movements from the Second World War onwards had an important role as an antidote to the destructive effects of the war and the associated extreme nationalism, fascism and other forms of totalitarianism.[33] Therefore, in this sort of interpretations the ECSC was a manifestation of a post-war West European peace settlement which resolved a specific economic problem, although its primary purposes were political.[34] Perhaps the most explicit argument in this vein has been put forth by Ole Wæver, according to whom the EU has evolved into a security actor whose most important function is to prevent Europe from returning to its own conflictual past.[35] Along these lines, as already implied, the big bang enlargement of 2004 is often justified by referring to its stabilizing effects and capacity to prevent conflict formation between formerly divided parts of Europe.[36] The same justification can be extended to the 2003 initiated European Neighbourhood Policy (ENP), which deals with co-operation with the enlarging EU's new neighbours in the east and south, many of whom are important energy transit or energy production sites.

The Narrative Context of the Energy Dialogue

The economic co-operation and politico-normative peace project narratives exist simultaneously and can even form parts of the same texts as seen in the case of the founding treaties of the EU. Policy-makers and analysts can make use of both in

---

32 Wiberg, 'Adaptive Patterns and Their Deep Roots', pp. 46–7; for the Baltic new members, see also Aalto, *European Union and the Making of a Wider Northern Europe* (London: Routledge, 2006), ch. 3, chs 5–7.

33 Walter Lipgens, *A History of European Integration, Vol I, 1945–47* (Oxford: Clarendon Press, 1982).

34 Dinan, *Europe Recast*, p. 321.

35 Ole Wæver, 'The EU as a Security Actor: Reflections from a Pessimistic Constructivist on Post-Sovereign Security Orders', in Morten Kelstrup and Michael C. Williams (eds), *International Relations and the Politics of European Integration: Power, Security and Community* (London: Routledge, 2000), p. 263.

36 Paolo Cecchini, Erik Jones and Jochen Lorentzen, 'Europe and the Concept of Enlargement', *Survival*, 43/1 (2001): pp. 155–65.

order to draw the wider political context into which they wish to situate the energy dialogue. But the way in which these grand stories are used, or how elements of them are combined, leads to very different possible and actualized orientations towards the energy dialogue – therefore their analytical separation. In addition, there are also some *counter-narratives*, as well as some less prominent narrative ingredients, or parts of stories, which can be used. The latter are best taken as *non-stories*, which lack the same narrative structure of having a beginning, middle part and an end.[37] Thus the non-stories are less comprehensive and coherent justifications for situating a particular policy position into the realm of socially accepted knowledge. Their proponents either rely on a critique of dominant narratives, or present new ideas that only partially challenge the more established and comprehensive variants. These entanglements within the wider narrative context of the energy dialogue give rise to several remarks on what sort of an issue and actor landscape we now have in European energy policy and what changes into these patterns may be about to ensue.

The Dominance of the Economic Co-operation Narrative

Compared to how the instituting of the ECSC and the EURATOM treaties combined the economic co-operation and political peace project narratives, it is striking that mainly the economic co-operation narrative seems to be in use for the promotion, justification and focusing of the present energy dialogue by the EU Commission and large businesses. One of the underlying conditions for this state of affairs is that in the 2000 Feira Council, the EU somewhat covertly agreed to decouple economic questions from the political ones in its Russia relations (see Chapter 4). The typical EU principle in dealing with its neighbours, political conditionality for economic co-operation, was thus relaxed in Russia's case. Yet, the EU Commission's Communication document of the progress during the first four years of the energy dialogue (2000–4) starts by linking up both to the political and economic narratives:

> The Energy dialogue launched at the summit between the European Union and the Russian Federation in October 2000 arose from the notion that *the European continent constitutes a broad geopolitical area linked culturally, historically and economically* and that the complementarity in terms of energy between the eastern and western parts of the continent should be developed in a sustainable way in the future.[38]

After this grandiose opening, the Commission's document proceeds to note the dialogue's pragmatic orientation in focusing on the 'resolution of problems'. The Commission grants that the dialogue offers 'wider prospects' going beyond narrow questions of energy trade, but sees them only extending to 'transport-related

37 Roe, *Narrative Policy Analysis*, pp. 3; 82.

38 European Commission, 'Communication from the Commission to the Council and the European Parliament: The Energy Dialogue between the European Union and the Russian Federation between 2000 and 2004', p. 2; emphasis added.

problems and to the environmental impact of the energy sector'.[39] The rest of the document is characteristic of the Commission's narrow and technical approach. The Commission's vision clearly falls short of comprehensively tackling the wider political context and its issue linkages going beyond mere technical ones. In this context, it is notable that since the Feira summit, EU–Russia economic relations have gathered new speed, with the Commission and leading member states dropping a lot of their former attention to Russia's democracy and human rights situation. This is not to deny that the EU and Russia have underwent several disputes, like in 2002 on Russia's transit to and from Kaliningrad after the EU's enlargement into Lithuania and Poland, and in 2004 the disagreement on extending the EU–Russia Partnership and Co-operation Agreement (PCA) and its stipulations on EU–Russia trade into the new CEE member states. As will be explained below, some of the new members, in particular Poland and partly also the Baltic states, have attempted to bring in a much more uncompromising stance into the EU's relations with Russia, thereby trying to re-politicize the increasingly instrumental character of the EU–Russia relationship.

The Commission's effort to stick with the economic co-operation narrative as far as the practical conduct of the energy dialogue is concerned, has played a part in the EU–Russia common economic space project becoming the best advancing area of mutual relations. Progress within the other common spaces – the common space for internal security, common space for external security, and the common space of research, education and culture – has been somewhat slower.[40]

The dominance of the economic co-operation narrative is understandable, given that the ECSC in terms of its substance also dealt with energy related questions, as did the EURATOM project. At the same time, it is striking how serious questions related to the EU's future viability as a political entity are now dealt with on a relatively narrow economic plane. The evolvement of the energy dialogue's agenda indicates a considerable concentration into the economic and relatively technical domains, even though the resolution of these questions can easily be linked to more explicitly political considerations (see also Chapters 3 and 4).

Functionalist Logic Leading to an Economics–Politics Linkage?

Despite the dominance of the economic co-operation narrative, functionalist logic suggests that the wider political context may strengthen as a result of increasing

---

39 Ibid.

40 Within the external security common space, crucial problems remain. Within the common space for internal security, a readmission agreement was signed in 2005, and visa procedures were speeded up and facilitated for students, close relatives and the disabled, with diplomats exempted altogether. Within the common space on education and culture, the EU and Russia have education and research exchange, and co-operate in research and development programmes. A further issue is the entry of several Russian universities to the Bologna process of harmonising higher education degrees across Europe. The establishment of a European Studies Institute in Moscow was agreed during 2005; see also Pami Aalto, 'The EU, Russia and the Problem of Community', in Hartmut Mayer and Henri Vogt (eds), *A Responsible Europe? Ethical Foundations of EU External Affairs* (Basingstoke: Palgrave, 2006), pp. 104–7.

economic energy co-operation. There is a debate whether functionalist logic holds in the case of the EU itself, but clearly the scope of EU integration has expanded over the decades, despite the fact that in some more recently added sectors co-operation remains thinner and more sporadic like in the Common Foreign and Security Policy (CFSP) and European Security and Defence Policy (ESDP).

Social constructivist theories of socialization suggest an argument related to functionalism: the parties are likely to develop some sort of common practices in the course of more institutionalization and meetings among policy-makers as they are discussing and in the process learning from each other. This could over the long run lead to the strengthening of a political understanding and co-operation between the EU and Russia. Nevertheless, for the moment, the case remains that after the Feira decision in 2000, EU–Russia relations have moved to a more instrumental direction, with both sides accentuating the economic side of the relationship on which it is easier, albeit not simple, to agree on than on the case of more explicitly value-laden questions for example related to who, and to a what extent, is involved in a normative project of peace building and overcoming of dividing lines in Europe.

The evident consequence of the dominance of the economic narrative thus is that the sector where EU–Russia integration is proceeding the fastest, remains ill equipped to contribute to the intensification of such more encompassing political relations. This feature is made worse when one keeps in mind that the Russian side has a long-standing aim of creating a 'Common European Home', as it was called during Gorbachev's era, or, as the more recent versions go, a 'Europe free of dividing lines', or a 'greater Europe' where Russia would also have a more integral role.[41] And for many EU member states – for example to the northern members Finland and Sweden, but also to big members like France, Germany, and to some extent, Italy – engaging Russia positively would constitute one of the best means of peace building in Europe.

Energy Dialogue Safe from Eurosceptics?

That the energy dialogue has so far resisted relatively well any pressures of confronting the wider political context explicitly, has also meant that its conduct has remained in few hands, and hence safe from any potential claims from the part of what can be called overt euro-sceptic NGOs and political movements. Their agenda has in recent years contributed to EU integration debates in many member states. The overt euro-sceptics do not subscribe to the economic co-operation or political peace project narratives, but rather rely on a *counter-narrative* of a European conspiracy.

This counter-narrative suggests that international capital and its embodiments such as the US as the most powerful state, institutions like the WTO, International Monetary Fund (IMF), etc., where international capitalists allegedly operate and shape the world, are responsible for starting and promoting EU integration in order to advance their own narrow interests. Moreover, these narrow capitalist circles with money-laden pockets are the ones reaping the benefits of member-states' contributions to the integration project. In more extreme versions, conspiracy

41 See Pami Aalto, 'The EU, Russia and the Problem of Community'.

theories are targeted at groups like the Jews and freemasons, who are counted among the ones exploiting the participant states and their potential. Although some versions of this counter-narrative seem extreme, its strongly euro-sceptic message should not be neglected. This is because it has followers in the member states, and is also represented by some euro-sceptics in the European Parliament. In short, the various versions of this narrative appeal to those with a sceptical stance towards the EU – especially in conditions of the Union having difficulties with its legitimacy and its Constitutional Treaty stalled in 2005. During the build-up to the 2004 enlargement, sentiments reminiscent of this narrative were also expressed by anti-EU movements in the accession states.[42]

The chief reason for the energy dialogue's remaining safe from any potential claims from the part of euro-sceptics is the fact that energy policy within the EU has, in the absence of the Constitutional Treaty, remained primarily in the hands of national governments. Although the EU Commission is a central party to the EU–Russia energy dialogue, functioning as a 'facilitator' in creating an institutional and a political framework for European and Russian businesses and political circles to meet each other,[43] there has been relatively limited wider politicization of the energy dialogue within the Union. In this fashion, the energy dialogue has remained an expert debate regardless of the fundamental questions of the present and future of European integration that it poses.

Energy Dialogue Linked to Other Forms of Identity Politics?

The energy dialogue has also remained relatively safe from other forms of identity politics by NGOs and the civil society in general in the contemporary Union. Apart from the coherent counter-narrative of the overt euro-sceptics, there is a less coherent *non-story* that is used in the negative sense so as to discredit other stories of European integration. It is best understood as a non-story because it lacks an idea of the beginnings of EU integration. Rather, it criticizes the stage EU integration arrived at during the 1990s. This includes the 1992 Maastricht treaty, the re-surfacing of federalist ideas around the turn of the millennium before the subsequent intergovernmentalist backlash, and the (stalled) project of designing a constitution for Europe, not even to mention the emergence of the single currency, and the CFSP and ESDP that have alarmed more nationalist-leaning observers. According to this argument, European integration is in danger of proceeding for its own narrow sake of Europe-making, failing to fulfil any real (or nationally defined) need of policy making. Here, the stage European integration has arrived at is not seen as a rational response to policy problems, but rather as an aim in and of itself, and one which has developed into a harmful direction during the recent history of the EU.[44]

---

42 See e.g. Evald Mikkel and Geoffrey Pridham, 'Clinching the "Return to Europe": the Referendums on EU Accession in Estonia and Latvia', *West European Politics*, 27 (2004): 716–48.

43 Debra Johnson, 'EU–Russia Energy Links', in Debra Johnson and Paul Robinson (eds), *Perspectives on EU–Russia Relations* (London: Routledge, 2005), p. 192.

44 See Aalto, *European Union and the Making of a Wider Northern Europe*, pp. 154–5.

With the EU continuing the effort of strengthening its role in European energy policy, it is possible to foresee it becoming increasingly targeted by groups subscribing to this sort of identity politics. For example, left-wing populist politicians can put all the blame on Brussels for marketizing energy policy decisions, even in cases where their own countries might adopt more or less the same measures nationally, or where member states might have themselves asked for a policy response from the EU. This is the mechanism by which the Union so often becomes cast in a negative light, when an alternative address for complaints would be at the national level. Effectively, the Union then ends up becoming a tool in national identity politics. This is why the EU Commission has in the new millennium so often demanded better political commitment from the member states to common policies and their development.

The fact that one of the main declared arguments for joining the EU for many CEE states in 2004 was 'returning to Europe' and demonstrating their belongingness to Europe, at the expense of the former association with the Soviet Union and its east European empire, makes these polities vulnerable to for this type of national identity politics in case EU membership fails to deliver when called upon for. The future of Europe's gas supplies from Russia is worth briefly taking up here.

As mentioned in the introduction to this book, 78 per cent of Russia's gas supplies go through Ukraine. In January 2006, Russia's Gazprom temporarily stopped supplying gas to Ukraine as a result of a dispute over transit fees, and for making Ukraine's own import gas payments market priced in place of the earlier reduced rates regime dating back from the Soviet era. Gazprom's decision had the effect of reducing gas flows to its European customers as well. Ukraine denied siphoning gas intended for European markets, but Hungary reported 40 per cent reduced gas supplies, Poland 38.5 per cent, and Croatia, Slovakia and France 30 per cent, with Germany and Italy also experiencing some setbacks.[45] The Ukraine episode made Poland particularly vocal in its attitudes towards the security of Russian supplies, and this was only aggravated with the ongoing project of building the Nord Stream gas pipeline through the Baltic Sea bypassing Ukraine and Poland altogether (see Chapter 5).

In such supply cut-offs and prospects of remaining outside major EU–Russia pipeline projects, the significance of EU membership can suddenly become cast in a totally different light. Especially this concerns countries like Poland who have relatively recently joined the Union. They declared loudly their European vocation in the process of joining. They have expectations of EU membership helping to alleviate some of their security problems, and voluntarily left behind bilateral trade ties and preferential gas supply arrangements with their former patron's successor state Russia. How Union membership can start look different than initially conceived

45 Ukraine claimed to be entitled to taking up to 15 per cent of the transit gas on the basis of the Russian–Ukrainian transit agreement, and claimed to be using gas it had bought from Turkmenistan. However, Russia denied supplying Turkmen gas over which it has control as a transit country; see *Helsingin Sanomat*, 2–3 January 2006.

of, can at the more theoretical level be discerned with the help of constructivist theory, which suggests that identities and interests are interlinked.[46]

When practical interests of gas supply are endangered, that can to an extent feed into the identity side. But importantly, the identity/interests linkages can also lead to a strengthened perception among old and new EU members of a shared risk associated with Russian supplies. This calls for concerted EU action, and on such a calculation Poland relied in the build-up to the EU–Russia summit held in Helsinki in November 2006. Poland insisted that any progress with a new EU–Russia treaty replacing the 1990s PCA must include Russia's ratification of the ECT treaty regulating energy relations in the pan-European area. Poland's position effectively blocked progress on a new treaty in the summit, and was conjoined by a short dispute on meat imports from the EU area, including Poland, to Russia. Security of supplies concerns of the type raised by Poland were widely prevalent among Union members. They also yielded stronger Europe-belongingness, but on a narrow EU basis of shared risks, as opposed to the EU Commission's above-quoted statement of the whole European continent constituting a 'broad geopolitical area linked culturally, historically and economically'.

It is possible to predict that national identity politics and anti-Russian sentiments like in the Polish case will interfere in the conduct of the energy dialogue in the future as well. The 2006 Russo–Ukrainian and the subsequent Polish episodes were major instances raising popular expectations of EU action in the energy sector.

Energy Dialogue Encounters Regional Environmental Claims

Finally, there is a diverse group of actors telling *non-stories* centring around the theme of the environment. These are not complete accounts of European integration in the first place, and therefore are in this connection best understood as non-stories. But due to their role in the energy dialogue, they definitely deserve to be taken up here.

The starting point for understanding these non-stories is how environmentalism is typically found only in advanced industrialized societies of the type of EU member states. The more radical end of environmentalists put forth a gloomy depiction of industrialization and of economic growth hitting the limits that are set to it by the earth's stock of natural resources and the capacity of its ecosystems to support industrial activity, of which the energy industry is part of. European integration can be portrayed as one instance of such a limits to growth prospect materializing, with the Union exhausting its own energy resources in the way it currently is. The less radical end of these accounts comes up in the form of argumentation promoting sustainable development and in the form of interventions pointing out solutions to the environmental problems that are emerging to the decision-makers' sight within the energy dialogue.[47]

---

46 E.g. Iver Neumann, 'Ringmar on Identity and War', *Cooperation and Conflict*, 32 (1997): p. 323.

47 For a more complete typology of environmentalist arguments, see John Dryzek, *The Politics of the Earth: Environmental Discourses* (Oxford: Oxford University Press, 1997), pp. 12–15.

In Finland, Norway and Sweden, three north European countries bordering Russia, as well as in the Baltic states, environmental concerns have arisen due to Russia's increasing oil shipments close to their fragile coastlines. In Finland and Norway, references to energy issues rarely pass without notice of the potential environmental side-effects, with governmental actors mostly going for the less radical environmentalist positions and some civil society actors resorting more clearly to the limits to growth type of arguments (see also Chapter 6). The Baltic states seconded these concerns for example during the process of drafting the second Northern Dimension Action Plan in 2003.[48] And remarkably, in the Commission's communication document on progress within the energy dialogue during 2000–4, these Nordic and Baltic concerns are acknowledged in several places. The Commission claims to have received assurances from Russia that controls on oil tankers loading in Russia will be reinforced and that Russia is co-operating with efforts to eliminate single-hull tankers. Increased rail transports will be considered.[49]

On the whole, the Commission's progress document manifests acknowledgement of the less radical end of the environmental arguments, suggesting several problem-solving measures and speaking of the energy dialogue as an 'instrument of sustainable development'. Russia's ratification of the Kyoto Protocol is mentioned. This was in fact a requirement set by the Union in order to support Russia's accession to the WTO. Yet, to put the environmental arguments in their proper context, another EU requirement in the same context was that Russia gradually moves away from its double-pricing of energy resources for domestic and foreign buyers, and introduces market principles into its energy policy. But we should not preclude the possibility that if EU member states were to continue to claim more power to themselves as they recently have done within the Union, and especially if some big members like Germany were to join this particular thread of the environmental arguments, the weight of environmentalism on the whole can increase quite crucially in the conduct of the energy dialogue.

Conclusion: On the Place of Material Factors in EU integration

In this chapter, I have suggested that the role the EU–Russia energy dialogue plays in the present and future of European integration is dependent on what wider narratives of European integration one subscribes to. We need words to describe policy problems and events, and to put them into their wider context in order to make them intelligible. In this manner the wider story contexts into which we situate the energy dialogue simply influences the way we see its relevance.

The economic co-operation narrative of European integration was found to dominate contemporary European debates on the energy dialogue, being promoted by the EU Commission and also many large member states, as well as businesses. Within the premises set by this narrative, the energy dialogue is an absolutely crucial

48 Aalto, *European Union and the Making of a Wider Northern Europe*, chs. 5–7.

49 European Commission, 'Communication from the Commission to the Council and the European Parliament: The Energy Dialogue between the European Union and the Russian Federation between 2000 and 2004', pp. 7–11.

tool for the Union and member states to try and manage the risks associated with their more or less voluntary energy dependence on Russia. The crucial nature of the dialogue has even helped the EU to give up on its former insistence on its relations with Russia being based on notions like democracy and human rights. At this plane the Union has only recently started to chart alternatives to a shared energy future with Russia as one of the preconditions for the whole Union continuing to co-operate economically. However, whether they turn out to be viable remains an open question.

The politico-normative peace project narrative was found to be less utilized in this context despite the simultaneous attachment of many member states to it and its character as major platform of statesmanship, as well as Europe-making by NGO and civil society actors. Situating the energy dialogue as a thread into this wider narrative would provide a scenario and a channel helping to build a greater Europe, which is a long-standing aim of Russian foreign policy, and which is also a declared target of the EU's neighbourhood policy. This opportunity is largely missed at the moment.

The dominance of the economic co-operation narrative as a justification to the EU–Russia energy dialogue is comprehensible though, as this narrative is a very powerful one. And in a world where great rivalry between alternative economic models – liberalism versus socialism – is mostly overcome, and which is increasingly aware of the role of the economy and 'market forces' in sustaining and conditioning political and social life, the weight of this narrative might even be in the increase. This means that the political integration project of the EU, which for many is incomparable to any other presently existing political entities or processes, cannot escape reliance on its material basis. Whilst this may be an unsurprising concluding remark, there is weightier corollary to it: the contemporary EU does not explicitly and seriously try to exploit the integration potential within the material policy platform of energy trade for the purpose of advancing its other, non-material integration goals like peace-building in the wider European area.

Of course, there is the calculation on the EU's part that opening up a wider political agenda and actor landscape into the energy dialogue might increase internal discord within the Union and scare away its business partners, and hence might endanger the current incremental and technical progress within the dialogue. This would leave the EU worse off in terms of protecting its material basis, without which it would remain ill equipped to come up with any explicitly political or normative goals on other levels and in other occasions. International politics, and especially great power relations, remains a crude game where one can only make a long-lasting impact with an adequate material basis. But the flip side here is that as far as EU–Russia relations are concerned, the EU is trying to do this by relying on functionalist logic almost blindly. Seen from this angle, unless we are blind followers of functionalism, widening the narrative base of the energy dialogue might be the start of a useful advice to the Prince.

Chapter 3

Energy Dialogue and the Future of Russia: Politics and Economics in the Struggle for Europe

Viatcheslav Morozov

Introduction[1]

This chapter unravels some of the wider political implications of the EU–Russia energy dialogue for Russia, and also expands the discussion into EU–Russia relations and European developments as a whole. The starting point is that the EU–Russia energy dialogue, and, more broadly speaking, Russia's role in the energy field globally, has become significantly politicized during the past few years. Even if some would consider this trend unwelcome and avoidable, a careful look at the range of issues discussed in connection with the energy dialogue leads to the predictable conclusion that the whole exercise is not – and indeed cannot be – limited to finding the most efficient ways of delivering energy from Russia to the EU.

The very rationale of the energy dialogue lies in the fact that in order to accommodate both parties' priorities in the field of energy, one needs to address a number of other problems, some of which are relatively technical in nature, while others are highly politicized. Moreover, a thorough investigation of the structural, technical and political premises and implications of the energy dialogue suggests that these are directly related to the very nature of both the EU and Russia as political projects. The way we handle questions of energy can significantly impact not only the 'energy security' of both parties, but also the entire political structure of Europe. Consequently, it is impossible to solve the most pressing issues on the energy dialogue's agenda without addressing the wider framework of EU–Russia relations and, in particular, the fundamentally political question of Europe's outer boundaries.

In order to account for these wider issues, first, the EU and Russia are in this chapter treated as *projects* whose nature and role are continually (re)defined in political processes and which both have a crucial significance for the way we define 'Europe'. The EU–Europe of Brussels-centred integration is substantially different

---

1 Earlier drafts of this chapter were presented at the Annual Conference of the Finnish Society for Russian and East European Studies, Tampere, 17–18 March 2005, and at the conference 'Post-Soviet In/Securities: Theory and Practice', at the Mershon Center, Ohio State University, 7–8 October 2005.

from the Russian Europe of sovereign states and spheres of influence. Second, there is always a temptation to describe the energy dialogue exclusively as part of a mere economic co-operation narrative, which emphasizes the role of pragmatic logic of economic integration as a key factor underlying the creation of a single Europe. This chapter assumes that there is, indeed, a certain purely 'economic' role for Russia to play in the context of the energy dialogue: it is expected to provide Western Europe stable energy supplies which at the same time must also be safe – in environmental and other terms. But the central question in this chapter is whether this 'economic' role depends on any 'political' (or 'cultural') conditions. If the answer is yes, then the whole story should also be told in terms of a politico-normative narrative of Europe as a peace project (for the narratives, see Chapter 2), as well as related to the Russian narratives of sovereignty and great power politics. This is indeed where my argumentation will lead us, and this tentative conclusion will have some far-reaching implications.

But before going into more detail, it should be noted that by no means do I want to reify the boundaries between the economic and the political, or cultural realms of social life. The approach in this chapter self-consciously derives from political science, and is concerned with human agency and the nature of decision-making under conditions of indeterminacy, regardless of the substance of those decisions. Therefore, I use the labels 'economic' and 'political' (henceforth without quotation marks) as they are used in political discourse. The aim is to demonstrate that the discussion on the issues which are considered 'technical' and 'non-political', and therefore often removed from the sphere of public decision-making into the realm of technocratic politics, may have much wider consequences than usually thought of. It can therefore be said that in a certain sense, this chapter sets out to test some premises of modernization theory, i.e. the existence of a linkage between economic, social, demographic and political development. At the same time the ambition is to show that the very reliance on modernization theory – implicit rather than explicit – makes some political choices more probable than others and limits the space for political action that is, *inter alia*, a pre-requisite for the EU and Russia meeting each other in a pan-European context.

In order to realize these aims, in the next two sections I will discuss the nature of the EU as a political project and how it comes up in the energy dialogue, before moving to a more detailed discussion of the role played by Russia. The concluding section then outlines the implications with regard to what role the energy dialogue plays in the future of the wider European area embracing both the EU and Russia.

The Post-Enlargement EU: Trapped in an Imperial Logic

The understanding of the European Union as a political project implies taking its future as far from certain. Apart from the unknowables related to the possible establishment of a Constitution for Europe, there are at least three different directions for European integration to proceed. Sometimes one cannot but make the conclusion that the Union is trying to go along all three ways.

First, the Union is trying to consolidate as a (super) nation-state with a Westphalian brand, implying a clear division between the inside and the outside and a single sovereign centre governing domestic and foreign policies. This trend is evident in the expansion of qualified majority voting, in the efforts to establish the ESDP, in the internal security policies, etc. Second, the Union is trying to ensure its external security by continuous expansion, which enables one to describe it as an empire. The Union's governance is in this model structured as a series of concentric circles, centred around Brussels and fading towards the margins. The best illustration of this model is the policy of EU enlargement, which forces the periphery to accept certain conditions in order to move closer to the core. Third, there is a neo-medieval model which allows for power to be dispersed with multiple regional centres competing against and/or complementing each other depending on the issue in question. This metaphor arguably was behind the Northern Dimension initiative, which aimed at opening up new political spaces into northern Europe stretching over, but not breaking, the existing boundaries of national and other communities.[2]

Each of the three images is relevant for the Union's relations with Russia. In accordance with the Westphalian logic, Russia is certainly perceived as a source of possible threats, which are to be warded off with such means as a visa regime, border controls, co-operation with the Russian authorities in fighting organized crime, financial aid for taming environmental threats emanating from Russia, etc. The imperial way of thinking, by contrast, presupposes that Russia is slowly but steadily involved in the concentric circles of European integration and offered incentives to move from the periphery to the core by fulfilling certain conditions, with a view of 'normalizing' or 'Europeanizing' the country and thus making it part of the continuous European political space with its centre in Brussels. Christopher S. Browning has demonstrated the paradoxical nature of the EU's attempts to combine these two different logics in its policy towards Kaliningrad.[3] The European Neighbourhood Policy is developed as a possible way out of this dead end after the 2004 enlargement, when it became clear that such new neighbours as Belarus, Moldova and Ukraine can hardly be expected to join the EU in the near future.[4] However, in the end Brussels failed to offer a radically new approach, and the same logic of conditionality continued to apply not only to Bulgaria and Romania, or Croatia and Macedonia, but even to those states which cannot hope to become members any time soon, and for whom the incentives for focusing their efforts on meeting the EU conditions in order to move closer to the core are therefore rather low.[5]

---

2 Christopher S. Browning, 'Westphalian, Imperial, Neomedieval: The Geopolitics of Europe and the Role of the North', in Christopher S. Browning (ed.), *Remaking Europe in the Margins: Northern Europe after the Enlargements* (Aldershot: Ashgate, 2005), pp. 85–101.

3 Christopher S. Browning, 'The Internal/External Security Paradox and the Reconstruction of Boundaries in the Baltic: The Case of Kaliningrad', *Alternatives*, 28/5 (2003): 545–581.

4 Fabrizio Tassinari, 'Security, Integration in the EU Neighbourhood: The Case for Regionalism', *CEPS Working Document*, no. 226, July 2005 <http://www.ceps.be>.

5 Pertti Joenniemi and Christopher S. Browning, 'Discourses on Centrality and Marginality: The European Neighbourhood Policy and Finnish Options of Europe-Making',

Although Russia officially is not part of the ENP, in practice the same logic of conditionality is applied to EU's relations with Moscow and to cross-border co-operation with the adjacent Russian regions. As Dmitry Danilov from the Institute of Europe of the Russian Academy of Sciences notes with disappointment, 'Brussels still seeks to apply its own integrationist logic to Russia, which implies a steady adoption [by Russia] of the EU's standards, norms and policy'.[6] The following section of this chapter demonstrates that the same logic of conditionality works in the case of the EU–Russia energy dialogue, although the specific nature of the energy sphere modifies its application to a significant degree. It will be argued, then, that this logic is doomed to failure when applied to Russia in general, and that its chances for success in the case of energy are not as good as in other areas. The neo-medieval logic is often suggested as the best way out of this impasse, but its application into EU–Russia relations, as shown in the final section, will require a significant transformation of the ways in which Europe is conceptualized both in Moscow and in Brussels.

EU Energy Dependency and Security of Supplies: the Role of Russia

One would not fall wide off the mark by arguing that the European logic of conditionality is implicitly based on the premises of modernization theory. The first official policy paper on the ENP maintained, for instance, that 'regional and subregional co-operation and integration … are preconditions for political stability, economic development and the reduction of poverty and social divisions'.[7] In general, at the core of the European Union's identity discourse lies a firm belief that by consistently and simultaneously widening and deepening integration, European states will be able to leave behind the centuries old legacy of mutual hostility.[8] Leaving the past behind by adopting the *acquis* has also been the key idea behind the EU's attempts to deal with the legacy of the Cold War. It is easy to note that these attempts are based on the idea of a linkage between the economic and the political which forms the essence of modernization theory.

It is hardly surprising therefore that this idea of an economics–politics linkage, lying so deep at the core of the EU as a political project, is also applied to the somewhat different agenda of the energy dialogue. Yet, in this case, the main driving force for the EU's involvement is of course not the concern for 'political stability' or the desire to overcome 'divisions in Europe', but rather the need to ensure long-term security of energy supply. The European Union's Green Paper on Energy Security,

---

paper presented in the VII ICCEES World Congress, Berlin, 25–30 July 2005.

6 Dmitrii Danilov, 'Dorozhnye karty, vedushchie v nikuda', *Nezavisimaya gazeta*, 24 May 2005.

7 European Commission, 'Wider Europe – Neighbourhood: A New Framework for Relations with our Eastern and Southern Neighbours', Communication from the Commission to the Council and the European Parliament, COM(2003) 104 final, 11 March 2003, p. 3.

8 Ole Wæver, 'The Temporal Structure of European Security Identity', paper presented in the International Studies Association Annual Convention, Honolulu, HI, March 2005, <http://www.isanet.org/archive.html>.

published in November 2000, maintains that geographic diversification of supply is highly desirable.[9] On the other hand, the paper describes a greater dependence on Russia for gas supplies as 'inevitable' in view of the fact that about one-third of world reserves are located in that country.[10] Noting that 'gas supplies from the Soviet Union, and then Russia, over the last 25 years is testimony to an exemplary stability', the document stresses the need for a long term strategy in the framework of a partnership with Russia.[11] The need to secure stable energy supplies has an obvious impact on the interplay between economics and politics.

According to some reports, an informal agreement was reached at the Feira European Council in 2000 to decouple the EU–Russia energy dialogue from political conditionality usually applicable to the Union's relations with Russia (see Chapter 4). However, this agreement seems to refer only to 'openly' political issues such as Chechnya or the freedom of the media. The strategic vision of the EU's relations with Russia in the energy sector, as expressed in the Commission's Communication of December 2004, is still firmly grounded in the belief that the only way to ensure stability of supplies on the part of Russia is to spread the principles of the EU internal market beyond the Union's borders.[12] The list of priorities in the first synthesis report on the EU–Russia energy dialogue included the reform of the Russian natural monopolies, improvement of investment climate, market opening, access of foreign companies to exploration, production and transportation of energy resources, security of transport networks (including transit), and improved energy efficiency. As for long term goals, special emphasis was laid on co-operation in the field of climate change (a major concern put forward in the 2000 Green Paper) and on increasing energy efficiency, in particular, by more extensive use of renewable energy sources.[13] But here one should note that many of these allegedly joint goals were already included in the ECT treaty signed in 1994, and in its transit protocol, which Russia signed but never ratified. In April 2004 the ECT was finally removed from the State Duma's agenda as 'flatly contradicting national interests of Russia' and 'being imposed on Russia' from the outside.[14]

---

9 European Commission, 'Towards a European Strategy for the Security of Energy Supply', Green Paper, COM(2000) 769 final, 29 November 2000, p. 41.

10 Russia's export capacity, measured as the difference between reserves and internal consumption, is greater than that of the Middle East, Africa, Central and South America combined: see Nodari Simonia, 'The West's Energy Security and the Role for Russia', *Russia in Global Affairs*, 2/3 (2004): 102–103.

11 European Commission, 'Towards a European Strategy for the Security of Energy Supply', p. 40.

12 European Commission, 'The Energy Dialogue between the European Union and the Russian Federation between 2000 and 2004, Communication from the Commission to the Council and the European Parliament, COM(2004) 777 final, 13 December 2004; cf. Chapter 4 of this volume.

13 'EU–Russia Energy dialogue', synthesis report presented by Russian Vice-Prime Minister Viktor Khristenko and European Commission Director-General François Lamoureux, Brussels/Moscow, September 2001.

14 Chairman of the Duma Committee on Energy, Transport and Communications Valery Yazev, as quoted by the ABN on 23 April 2004; see <http://www.gazo.ru/ru/main/news/

The logic behind all these measures is clear: the EU is trying to put up with the fact that it will depend on energy imports for decades to come, and to develop a stable relationship with Russia as arguably the most reliable supplier. At the same time, it is assumed that the reliability of Russian supplies would increase with the de-monopolization and internationalization of its energy sector. Growing number of actors in energy production and export would, according to this way of reasoning, mean fewer opportunities for creating price cartels and for the intervention of non-economic factors – such as the state putting pressure on energy companies to restrict supplies or increase prices for the sake of geopolitical goals. Higher foreign direct investment (FDI) in the energy sector is likely not only to make oil and gas companies more responsive to EU concerns, but also to generate additional flow of capital into the exploration and development of new reserves. Increased energy efficiency within the domestic economy can reduce the growth in domestic demand for energy, which will mean more oil and gas available for export. Finally, by developing and ensuring equal access to transport infrastructure the EU will further diversify its supplies geographically, as this will make energy resources of Central Asia and the Caspian available on the European market through Russian pipelines.

It is hardly possible to say that all these considerations are irrelevant for the EU, which is striving to overcome the relative inefficiency of its economy as compared to the US and the fastest-growing Asian countries, which, *inter alia*, requires curbing the growing energy spending. At the same time, there is a number of arguments against what Susanne Peters calls 'putting all eggs into the Russian basket',[15] including the disparity between production capacity and export commitments. As Juhani Laurila bluntly put it, '[s]ubstantial financing will be necessary to mobilize Russian energy resources, and their abundance is illusory'.[16] Thus, in spite of the much-quoted fact that the Soviet Union/Russia has been the most reliable supplier of energy for the last 30 years, some doubts persist with regard to its ability to remain equally faithful in the future (see also Chapter 2).

Some of the concerns associated with the Union's reliance on Russia boil down to technical issues which are already addressed by the EU. Russia could significantly increase its energy exports by freeing additional resources in the domestic market, first of all through energy saving measures. An encouraging sign is that the energy efficiency of the Russian economy, which is rather low in comparative terms, is increasing faster than in the post-industrial west.[17] The issue of energy efficiency figures prominently among the priorities of the Energy Strategy approved by the

---

news_current.shtml?2004/04//420.html>.

15 Susanne Peters, 'Courting Future Resource Conflict: The Shortcomings of Western Response Strategies to New Energy Vulnerabilities', *Energy Exploration & Exploitation*, 23/1 (2003): 36.

16 Juhani Laurila, 'Transit Transport between the European Union and Russia in Light of Russian Geopolitics and Economics', *Emerging Markets Finance and Trade*, 39/5 (2003): 29.

17 Alexander Arbatov, Vladimir Feygin and Victor Smirnov, 'Unrelenting Oil Addiction', *Russia in Global Affairs*, 3/2 (2005): 147.

Russian Government in 2003.[18] Energy dialogue addresses this issue by promoting Tacis-sponsored pilot projects in Archangelsk, Astrakhan and Kaliningrad.[19]

One may thus argue that the EU has addressed the issue of energy savings in its dealings with Russia (see also below). Even if one may recommend greater effort in this direction, strategically the policy is the right one. There are, however, spheres where one may locate even more crucial problems related to the long-term goals of the energy dialogue and even perhaps to the overall vision of future relations between Russia and the EU. The political implications of the seemingly technical issues of gas prices and pipeline routes are simply too weighty to be ignored. The starting point here is the problem of foreign investment in the energy sector. Both sides seem to agree on the importance of FDI for the development of Russia's production and export capacity. Russia puts more emphasis on the need for investment as such, while the EU tends to underscore the importance of improving the investment climate. This discrepancy may look a mere question of emphasis, but if one considers the approach of each party more carefully, and in view of the recent political developments in Russia, one may come to the conclusion that the disagreement is more serious than at first appears. In the final analysis, the attitude to foreign investment depends on both parties' images of each other and of their respective place in the new Europe. At stake is the profoundly political question of where the community of Europe ends.

Investment and Interdependence vs. Strategic Control and Geopolitics: the Russian Dilemma

Some authors have argued that '[a] long-term steady and growing energy partnership with Europe will give Russia a stake in the EU's future, provide a means to transform Russia's economy, and perhaps begin to foster a sense of connection with a steadily expanding definition of Europe, if not one of association with the EU as a security community'.[20] Russia's Vice-Prime Minister Viktor Khristenko who is also Russia's chief minister responsible for the energy dialogue, foresees that deeper co-operation with the EU can help Russia to introduce 'new standards of quality and governance'.[21] Dmitry Danilov maintains that a new strategic vision to the relationship could actualize 'the potential of Russia's social and economic development and democratic modernization'.[22] All that can be perfectly true, but

18 'Energeticheskaya strategiya Rossii na period to 2020 goda', utverzhdena rasporiazheniem Pravitel'stva Rossiiskoi Federatsii ot 28 avgusta 2003 g. no. 1234–r,', p. 16ff, <http://www.mte.gov.ru/files/103/1354.strategy.pdf>.

19 'S pritselom na prochnoe partnerstvo', *Dialogue*, special issue for the Russia–EU Summit, St. Petersburg, May–June 2003.

20 Amy Myers Jaffe and Robert A. Manning, 'Russia, Energy and the West', *Survival*, 43/2 (2001): 146.

21 Viktor Khristenko, 'Nuzhna li nam integratsia?', *Rossia v globalnoi politike*, 2/1 (2004): 78.

22 Danilov, 'Dorozhnye karty, vedushchie v nikuda'; see also Arbatov et al., 'Unrelenting Oil Addiction', pp. 152–155.

the plausibility of these arguments hinges on a number of factors, ranging from pragmatic calculations to identity politics.

The incentives for Russia to accept the *acquis* as the only basis for the development of an integrated energy market and as a condition for its possible rapprochement with the EU eventually depend on the possibility to create a shared identity between Russia and Western Europe. So far, the impression is that both parties see each other as two separate entities that are destined by geopolitics to compete against each other, with this competition at times turning hostile. Unlike in the case of CEE countries which do not see their future outside the European structures, Russia perceives itself as a centre of its own and therefore advances a different agenda. Very much like the EU, Russia as a political project is being increasingly conceived of as an empire with the centre in Moscow and the sphere of influence stretching all over the post-Soviet space.[23] Contrary to the EU's logic of conditionality, Danilov notes, 'Moscow prefers to speak about co-operation on equal terms, about a two-way street'.[24] Being confronted with the tough position of the EU bureaucracy, which refuses to consider any proposal about a 'special relationship' with Russia or any other way of recognizing Russia's role as an independent power in the new Europe, Moscow reacts by using all means – real or imaginary – of insisting on its independent position and influencing the pan-European developments.

The field of energy provides plenty of opportunities for this type of political games. The energy dialogue is unfolding under the conditions of scarcity: energy becomes more and more valuable resource on the global scale. Policy-makers and economic agents all over the world become increasingly aware of the fact that oil and gas are limited and exhaustible resources, which makes competition for them more and more acute. Besides, Russia cannot significantly increase production in the near future because of transportation bottlenecks and the lack of easily accessible new reserves (see Chapters 2 and 8). There are alternative buyers who can, at least potentially, offer significant rewards for redirecting the transport infrastructure away from Western Europe. These rewards go far beyond economic factors such as investment. To start with, the US, in spite of the recent tensions, still largely treats Russia as one of the major partners in the antiterrorist coalition. Japan can potentially offer some concessions for the territorial settlement of the long-standing Russo–Japanese border dispute on the Kurile Islands. China is a key member of the Shanghai Co-operation Organization and part of the grand geopolitical designs cherished by the Russian political elite. Diversification of exports is one of the priorities of Russian external energy policy as defined in the Energy Strategy,[25] and a major step in this direction was made with the signature of the memorandum on the construction of gas pipelines from Russia to China during President Putin's visit

23 For a more thorough discussion of Russia's new imperialism, see Viatcheslav Morozov, 'New Borderlines in a United Europe: Democracy, Imperialism and the Copenhagen Criteria', in *Russia's North West and the European Union: a Playground for Innovations* (Nizhny Novgorod: R-US Expert Transit and Danish Institute for International Studies, 2005), pp. 74–84.

24 Danilov, 'Dorozhnye karty, vedushchie v nikuda'.

25 'Energeticheskaya strategiya', p. 41.

in March 2006.[26] It is obvious that Russia is trying to increase its action capacity in the framework of the energy dialogue with the EU. This, however, involves the need to control the domestic energy sector so that the government can use energy as a foreign policy tool.

It is well known that the Russian Government, at least in the economic field, can hardly be described as a unitary actor with a solid and coherent agenda (cf. Chapter 8). Up until late 2005 the cabinet, as well as some of the key ministries, had been divided into liberals and proponents of economic nationalism, and the whole history of Russian reforms, including the most recent ones, abounded with contradictory steps and declarations. Nevertheless, towards the middle of President Putin's second term in office, it became increasingly evident that the liberal reform projects made no headway, whereas economic nationalist agenda was gaining prominence. Taking stock of the political evolution of the Federal Government in 2005, the leading liberal newspaper concluded that 'bogged down in bureaucratic intrigues and having abandoned their conflicting opinions', liberal cabinet members lost their influence and resigned to a distinctly nationalist turn in economic policy.[27]

It was in the energy sector where this turn actually was initiated. The plans to reform Gazprom in order to introduce competition into the highly monopolized gas market were largely abandoned as early as in 2004. As stated by President Putin in a meeting with German Chancellor Gerhard Schröder, '[w]e intend to preserve state control over the gas pipeline system and we will not divide Gazprom'.[28] Russia–EU agreement on Russia's WTO entry allowed the export monopoly of Gazprom to be preserved, while the only concessions on Russia's part consisted in the pledge to gradually raise internal gas prices and to guarantee trunk pipeline access to all internal producers. In the electricity sector, the discussions on a reform look more up to the point, but they largely remain to be put into practice. In the oil sector, in spite of all remaining uncertainty as to the actual motives and driving forces of the Russian Government's policy, one thing is clear: the state is determined to strengthen its control over strategic energy production and export. This was well attested to in the development of the criminal case against the top management of the oil giant Yukos, including its politically ambitious former leader Mikhail Khodorkovsky, during 2004–5, and in the follow-up to the whole episode thereafter. The 'Yukos affair' resulted not only in Khodorkovsky being sentenced for tax evasion, embezzlement and other crimes, but also in the effective nationalization of the most important part of Yukos, Yuganskneftegaz. At the December 2004 auction it was acquired by the state-controlled oil company Rosneft, thus creating a second energy giant directed from the Kremlin, to complement Gazprom's gas monopoly. This trend continued and spread into other industries, such as engineering (the purchase of Silovye

---

26 Vladimir Kuzmin, 'God Rossii po kitaiskomu kalendariu', *Rossiiskaya gazeta*, 22 March 2006.

27 Dmitry Butrin, 'Pravitel'stvo natsional'nogo proekta', *Kommersant*, 30 December 2005.

28 Vladimir Putin, 'Vystuplenie na torzhestvennom sobranii, posviashchennom 10–letiu OAO Gazprom', 14 February 2006, <http://www.kremlin.ru/text/appears/2003/02/29774.shtml>.

Mashiny by the Unified Energy Systems) and car manufacturing (the takeover of the AvtoVAZ by the Rosoboroneksport defence holding).

As maintained by *Kommersant*, the essence of the government's industrial policy from 2005 on 'consists of the distribution of spheres of business influence between the key formal and informal power structures and of supporting national security by means of squeezing out of the market all "suspect" (from the bureaucrats' viewpoint) investors'.[29] One may argue at length over the economic effectiveness of this approach, as well as over the means used to establish control over the privately owned companies, but it is obvious that increasing state pressure is hardly compatible with the need for new FDI. Even investors who are in principle interested in dividends alone and do not envisage establishing control over companies whose stock they acquire, are probably unlikely to risk their money by investing into a sector heavily controlled by the government, which can in the end force private companies to act against their economic interest.

The state consolidated its grip over the 'strategic resources' in 2006. In October, when the protracted process of selecting a foreign partner for Gazprom to develop the Shtokman gas field in the Barents Sea was nearing its end, the Russian gas monopoly suddenly declared that it would go on with this project on its own, with no international partners involved. The bitter disappointment of the would-be participants – the Norwegian Statoil and Norsk Hydro, the French Total and the US Chevron and ConocoPhillips – was relieved during the following month, when the top Russian officials, including President Putin, explained that foreign partners would be brought in, albeit 'in a different format', with Gazprom as the only licence holder for Shtokman's 3.7 trillion cubic metres of gas (see also Chapter 6).[30] As the Shtokman case unfolded, Sakhalin Energy, a company owned jointly by the Royal Dutch Shell of Britain and the Netherlands, and the Japanese Mitsui and Mitsubishi, was accused by the Russian authorities of violating the environmental norms in developing the major Sakhalin–2 project on the shelf of the Sakhalin Island in the Russian Far East. Characteristically, the crisis was resolved in December 2006 not by an environmental clean-up, but instead by Gazprom buying 50 per cent of shares of the international participants to the Sakhalin–2 project and thus effectively establishing control over its 600 billion tons of oil and 700 billion cubic metres of natural gas.[31] The Russian-British TNK BP was at the same time under similar pressure in relation to its Kovykta oil field in east Siberia, and has repeatedly declared its preparedness to cede control to Gazprom. However, it is the sea shelf oil and gas resources, such as Shtokman or Sakhalin–2, which seem to attract the Federal Government's most attention – to the extent that it has promised to introduce a law banning foreign companies from owning shelf fields, and may also be thinking

29 Renata Yambaeva, 'RAO "Gosprom"', *Kommersant*, 30 December 2005.

30 'Zapad dopustiat k Shtokmanu "na novykh usloviyakh"', BBCRussian.com, 8 December 2006, <http://news.bbc.co.uk/hi/russian/russia/newsid_6162000/6162197.stm>.

31 '"Gazprom" kupil kontrol nad "Sakhalinom–2"', BBCRussian.com, 21 December 2006, <http://news.bbc.co.uk/hi/russian/russia/newsid_6201000/6201705.stm#>.

to oust Russian private companies from such projects, thus concentrating all these resources for the state owned Gazprom and Rosneft.[32]

Increasing involvement by the state into the management of the energy sector cannot but reduce Russia's credibility as an energy supplier to the EU. One can, of course, dream up a situation in which the Russian Government will use its leverage over the energy sector in order to induce Russian companies to cater for the urgent needs of their European customers, especially in the case of a major price surge or supply crisis, or both. The former German Chancellor Schröder, for instance, presented his personal relationship with President Putin as a guarantee for the stability of energy supplies (see Chapter 5). The long-term record of stable Soviet and Russian energy supplies to the west European markets used to be a strong argument in favour of the latter point, but the more recent developments make the opposite scenario at least equally plausible.

In the early to mid–1990s, Russia used natural gas supply cuts to Ukraine to advance such demands as Ukraine to cede full control over the Black Sea fleet to Russia or join the customs union with Russia and several other states of the Commonwealth of Independent States (CIS).[33] Even before the 'gas war' of December 2005–January 2006 the danger of such manipulation was recognized both in Brussels and Washington: as Nikolai Zlobin put it, 'the US is not interested in the "energy switch" becoming the key and, most importantly, unpredictable element of Russia's foreign policy toward former Soviet republics and other countries'.[34] Former US ambassador to Lithuania and consultant to the Williams energy company Keith C. Smith went as far as to say that '[t]he US government and the EU should stop and reconsider the costs … of their rush to secure additional oil and gas supplies from Russia', and 'collaborate … to counter Russian energy policies that threaten the consolidation of democracy and free markets in Poland, Ukraine, Estonia, Latvia, and Lithuania'.[35]

This energy geopolitics *á la Russe* culminated in the dispute with Ukraine over the price of Russian natural gas at the turn of 2005–6. There are grounds to believe that legally speaking, Gazprom was fully entitled to demand a higher price[36] – at least the fact that Kiev did not bring the case before the Stockholm arbitration court indirectly supports this point. Yet, the reputation of Gazprom as a shady dealer and a political actor in the hands of the Russian Government rather than an independent market-oriented supplier, combined with the extremely poor performance of the Russian diplomacy during the Ukrainian orange revolution a year earlier, as well as

32 'Inostrantsev ne pustiat na Rossiiskii shelf', BBCRussian.com, 25 January 2007, <http://news.bbc.co.uk/hi/russian/business/newsid_6299000/6299677.stm>.

33 Keith C. Smith, *Russian Energy Politics in the Baltics, Poland, and Ukraine: A New Stealth Imperialism?* (Washington: Center for Strategic and International Studies, 2004), p. 47.

34 Nikolai Zlobin, 'Limited Possibilities and Possible Limitations: Russia and the U.S.: What's Next?', *Russia in Global Affairs*, 3/1 (2005): 219.

35 Smith, *Russian Energy Politics*, pp. 75–76; see also Keith C. Smith, 'Defuse Russia's Energy Weapon', *International Herald Tribune*, 17 January 2006.

36 Nikolai Sokov, 'Alternative Interpretation of the Russian-Ukrainian Gas Crisis', *PONARS Policy Memo*, no. 404, p. 2 <http://www.csis.org/media/csis/pubs/pm_0404.pdf>.

the general concern over the fate of Russian political and economic reform, led to a situation where the Ukrainian position was in advance accepted internationally as a legitimate standout against what was perceived as Russian imperialist pressure. At the same time, even if one may argue that Kiev won in terms of international public opinion, both countries may be said to have lost in the resulting deal, which created an absolutely non-transparent scheme for gas supplies to Ukraine, with the bulk of the profit going to RosUkrEnergo – a company whose owners prefer to remain unidentified.[37] Thus, while the declared initial objective of both parties was to make their trade in gas more transparent and market-based, the conflict made the situation only worse in those terms (see also Chapter 4).

It seems that during the recent years nearly any crisis in Russia's relations with its neighbours has invariably led to the deployment of the energy weapon, which sometimes also affects EU consumers. The chronic tension between Russia and the Baltic states was a major factor underlying the decision to build the Baltic Oil Pipeline System bypassing the Baltic ports and Finland (see also Chapters 7 and 8). The cost of the first two phases of the project was estimated at 3–5 billion US dollars, whereas a similar extension of transport capacity could have been attained through investing ten times less in the development of the transit route through Ventspils, Latvia. In addition, the latter option could have been supported by the World Bank, the European Bank for Reconstruction and Development (EBRD), and Tacis.[38] Additional evidence in favour of the argument that economic efficiency was in this case sacrificed to geopolitical concerns is that Russia has foregone some of the profits which could have resulted from the high oil prices by never resuming pumping oil to Ventspils, even while the BOPS was unable to transport all the oil available for export.

Moscow's persistence in advancing the Nord Stream gas pipeline project is also to a large extent explained by the desire to decrease Russia's dependence on the transit states (first of all, Ukraine, Belarus and Poland) and thus perhaps to acquire an additional tool for pressing them for geopolitical ends (see also Chapter 5).[39] Moreover, the acute crisis in Russia's relations with Georgia in autumn 2006 resulted in the beginning of the construction of a gas pipeline from Russia to the breakaway Georgian autonomy of South Ossetia, which would, in the words of a Russian official, prevent Georgia in the future from 'cutting off energy supplies vitally important for South Ossetia's inhabitants.'[40] Last but not least, the dispute with Belarus over energy prices, transit tariffs and – in the background – the future of the union between the two countries resulted in a temporary interruption of oil supplies through Belarus to Western Europe in the beginning of 2007. This development caused a bitter reaction

37 Even President Putin declares that he does not know who stands behind this company: Vladimir Putin, 'Interview to the Spanish Media', 7 February 2006, <http://president.kremlin.ru/eng/text/speeches/2006/02/07/2343_type82916_101277.shtml>.

38 Laurila, p. 46, 53.

39 Smith, *Russian Energy Politics*, pp. 17–18, 45–46.

40 'Nachalos stroitelstvo gazoprovoda v Yuzhnuyu Osetiyu', *Vzgliad*, 27 October 2006, <http://www.vz.ru/news/2006/10/27/54556.html>.

on the part of EU member states and provided Brussels with yet another argument in favour of an energy policy reform.[41]

The developments of the recent years, in particular since the Russo–Ukrainian 'gas war' demonstrate that Russia's monopolization of the energy sector is a source of anxiety for Russia's neighbours in light of the linkage between democracy and security that is so firmly established in western discourse: undemocratic states are expected to threaten global security and stability. Democracy, in turn, can fall victim of economic monopolization. The takeover of Vladimir Gusinsky's Media-Most holding by Gazprom in 2002 sets a very disturbing precedent of using resources accumulated in the energy sector to establish state control over independent media. But after all, one does not even have to prove that Russia is likely to use the monopolized energy sector to press its neighbours. Given the burden of history in mutual relations and the lack of transparency in corporate management, it is clear that a mere awareness of monopolization provides an additional argument against Russian transit and investment in the transport infrastructure of the CEE countries. This argument also seems to have currency far beyond the area.[42]

Even if this analysis provides ample evidence against 'putting all eggs into the Russian basket', it would, however, be equally unwise to put all the blame on the Russians. Rather, this analysis supports the argument that the seemingly technical issues within the energy dialogue are firmly embedded into a wider problematic of Russia–EU relations and relate to the fundamental question of belonging and exclusion. Accordingly, before one can raise the question of responsibility, one first has to discuss the alternatives available to each party.

It seems that in the case of Russia, a country rich in energy but squeezed between its imperial past and the lack of political and institutional resources for jumping right into the brilliant future, the alternatives are few, if any. By insisting on the principle of conditionality, the EU is leaving Russia with little choice – either it has to give up its self-image of an independent sovereign power and to integrate into the single market, with no chance whatsoever to have a say over the way the latter is regulated – or it has to consolidate political control over its energy sector and try to use it with a view of at least partially regaining its position as a great power. It is hardly surprising that Russian policy-makers choose the latter option. One may call their logic flawed, but it is probably no more flawed than that of the EU bureaucracy, which unwittingly relies upon the most simplistic version of the modernization theory, being unable to fully grasp how crucial Russia's self-understanding as a strong sovereign nation is for itself. This self-understanding is by no means limited to the 'purely' political matters: President Putin's statement in the December 2005 Security Council meeting that Russia should become a leader of the world's energy market,[43] and Russia's

41 See e.g. Aambrose Evans-Pritchard, 'Brussels Uses Energy Row to Push Tireless Federalist Agenda', *The Daily Telegraph*, 12 January 2007.

42 The most comprehensive assertion of this argument can be found in Smith, *Russian Energy Politics*.

43 Vladimir Putin, 'Opening Address at the Security Council Session on Russia's Role in Guaranteeing International Energy Security', 22 December 2005, <http://president.kremlin.ru/eng/speeches/2005/12/22/2222_type84779_99439.shtml>.

setting of energy security as the main theme of its G8 presidency[44] illustrate the key significance of energy for Russia's new identity. Secondly, and most importantly, any discursive transformation which could make the Russians more receptive to the appeal of the European integration project would require some sort of a reciprocal move on the part of west Europeans. As always, it takes two to tango.[45]

Costs and Limits of the Russian Strategy: the Energy Dialogue and the Future of Europe

At the same time as the Russian party is attempting to establish strategic control over its energy resources against the preferences of the EU, it is remarkable that the energy dialogue, unlike many other joint projects between Russia and the European Union, continues to function and even reportedly brings some concrete results. There is, however, a cost – at least for the EU. There are some crucial elements in the initial design of the dialogue that are almost completely missing in the current process.

As pointed out above, some progress has been achieved on energy saving projects, but one can hardly expect the pilot projects to become widespread practice without making this an issue in the public debate. The simple reason is that energy saving is as much a cultural as a technological issue. At the moment, the matter does not get any publicity in Russia, while both the Russian Government and the EU appear to ignore this aspect of the problem. Liberalization and legal approximation with the EU, another key element of the energy dialogue as it was initially planned, are certainly not in the interest of Russian monopolies, and it is therefore understandable why Moscow is trying to avoid a pointed dialogue with the EU on these issues. What is not as obvious is whether, in the long run, this strategy will benefit the Russian economy and the people of Russia. Ignoring the broader political dimension of such 'technical' and 'purely economic' issues in the framework of the energy dialogue is in itself a political decision which has serious consequences for the future of Russia and Europe as a whole. In general, what seems to be almost completely disregarded in the process is the possibility to use the energy dialogue as a means of promoting further reform in Russia and in this way bringing it closer to the EU. What is in question here is not the presumed need for Russia to adopt certain practices whose legitimacy is defined in Brussels, in order to 'integrate' into the EU–Europe (later

44 See in particular Putin's article published widely in the world media: Vladimir Putin, 'The Upcoming G8 Summit in St. Petersburg: Challenges, Opportunities, and Responsibility', 1 March 2006, <http://president.kremlin.ru/eng/speeches/2006/03/01/1152_type82914_102507.shtml>. The expert opinion about this decision has been for the most part negative – see e.g. Pavel K. Baev, 'Chairing the G8. Russian Energy and Great Power Aspirations', *PONARS Policy Memo*, no. 382 (2005), <http://www.csis.org/media/csis/pubs/pm_0382.pdf>; Vladislav Inozemtsev, 'Ne vpolne udachnyi vybor', *Nezavisimaya gazeta*, 25 January 2006; Leonard L. Coburn, 'God energeticheskoi nebezopasnosti', *Kommersant*, 24 March 2006.

45 Cf. Pami Aalto, 'EU, Russia and the Problem of Community', in Hartmut Mayer and Henri Vogt (eds), *A Responsible Europe? Ethical Foundations of EU External Affairs* (Basingstoke: Palgrave, 2006).

rather than sooner), but the lost opportunities for Russia's own economic and political development, for which co-operation with the EU can be a resource.

However hard the Russian Government tries to increase its action capacity in the field of the energy dialogue, this space will in the future remain necessarily limited for various structural reasons. One can hardly escape the fact that the EU is by far Russia's main trading partner, with more than half of the Russian exports going to the single European market. It is true that the lion's share of these exports consists of raw materials, including energy, for which Western Europe so far has no substitutes. But this huge share nevertheless provides Brussels with a chance to counter any unfriendly measure which might hypothetically be taken by Moscow in the energy field. Furthermore, Russia's position as the most important supplier of energy to EU member states and other countries of the region cannot be taken for granted in the longer run, since there are alternative producers eager to access the European markets.

For example, regardless of Russia's determined efforts to obstruct the realization of the Baku–Tbilisi–Ceyhan oil pipeline project from Azerbaijan to Turkey, the project was successfully completed by mid–2006, and even expanded to include Kazakhstan.[46] Armenia, frustrated by the price hike for Russian natural gas at the beginning of 2006, decided to speed up the construction of a gas pipeline from Iran. Even though Russia offered to compensate Armenia for expensive gas by supplying it with cheap weapons,[47] it is likely that Iranian suppliers will soon compete with Gazprom in this market. Competition may also come from quite unexpected directions. Every fourth new barrel of oil in the coming years will be produced in West Africa, where established players on the global energy markets (like Nigeria) and newcomers (Chad, Mauritania) alike are investing in infrastructure in order to increase their oil exports to Western Europe, East Asia and the US.[48] In spite of the fact that many countries of that region are often associated with political instability, taken as a whole, these oil supply routes can provide some extra leeway for European buyers.

The impact of the Russian–Ukrainian gas crisis on the EU's energy policies is also clearly visible at this plane. On surface, Brussels has accepted the Russian proposals about energy security as a new conceptual framework which has supplemented, and sometimes even replaced, the traditional EU emphasis of energy efficiency – a

---

46 See Radio Free Europe/Radio Liberty Newsline, part I, vol. 9, no. 147, 5 August 2005, and no. 127, 14 July 2006.

47 Ara Tatevosian and Aleksandr Reutov, 'Armeniu ubedili oruzhiem', *Kommersant*, 24 March 2006.

48 Anthony H. Cordesman and Khalid R. Al-Rodhan, *The Changing Risks in Global Oil Supply and Demand: Crisis or Evolving Solutions?* (Washington: Centre for Strategic and International Studies, 2005), p. 55, <http://www.csis.org/media/csis/pubs/050930_globaloilrisks.pdf>. One of the projects currently contemplated by the investors is the Trans-Saharan pipeline, which, even being a fairly long time off, may pose a huge challenge to the Russian energy exporters. For more information on alternative sources of energy supplies to Europe, see Ostap Karmodi, 'Ne truboi edinoi', *Moskovskie novosti*, 13 January 2006; Mikhail Zygar', Igor Fediukin and Natalia Grib, 'Kak Evropa budet spasat'sia ot Rossii', *Kommersant-Vlast'*, 16 January 2006.

term which obviously has much less direct political currency. However, contrary to Moscow's expectations, Russia more often figures as a source of insecurity than as a provider of stability on the energy market. As *Kommersant-Vlast'* put it in 2006, '[w]hile up until now Vladimir Putin has offered to ensure energy security of Europe with the help of Russia, as of the beginning of the New Year the Europeans are desperately trying to secure themselves from Russia'.[49] In its March 2006 Green Paper, the European Commission explicitly sets the task of establishing the EU as an international actor in the energy field. The paper advocates 'a coherent external energy policy' aimed at securing 'independent gas pipeline supplies from the Caspian region, North Africa and the Middle East into the heart of the EU', as well as creating new liquefied natural gas terminals and 'Central European oil pipelines aiming at facilitating Caspian oil supplies to the EU through Ukraine, Romania and Bulgaria'. The outline of external energy policy towards Russia includes such keywords as 'predictability', 'third party access to pipelines', 'rapid ratification by Russia of the Energy Charter Treaty and conclusion of the negotiations on the Transit Protocol' (!).[50] A key outcome of Moscow's attempts to use energy as a policy tool thus is the bolstering of the EU as an energy policy actor,[51] with the aim of countering Russia's monopolistic measures and promoting the diversification of supplies.

Another future-related limitation for Russia's energy-based geopolitics is less visible but more radical: oil and gas stocks will sooner or later come to an end, while the high price of hydrocarbons stimulates the development of new technologies which will sooner or later provide an economically sustainable alternative to fossil fuels. Besides, nuclear energy is being rehabilitated after decades of the 'Chernobyl syndrome'. Finland launched its fifth reactor in 2005; Germany is considering a reversal of the decision to close down its nuclear power plants, and a similar mood is spreading throughout Europe,[52] including Russia itself.[53] Thus, precious time is wasted. Instead of using the EU as a possible source of new technologies, Russia is overexploiting its antiquated energy sector to increase its geopolitical profile.

Apart from the future perspectives, most frustrating are the wider political implications of the current modality of the overall relations between Russia and

---

49 Zygar' et al., 'Kak Evropa budet spasat'sia ot Rossii'.

50 European Commission, 'A European Strategy for Sustainable, Competitive and Secure Energy', Green Paper, COM(2006) 105 final, 8 March 2006, p. 15. Later on, the EU suggested it could lift the demand for Moscow to ratify the ECT if its principles, as well as those of the Transit Protocol, are integrated into the new comprehensive agreement between the EU and Russia: European Commission, 'An Energy Policy for Europe', Communication from the Commission to the European Council and the European Parliament, COM(2007) 1 final, 10 January 2007, p. 24.

51 Cf. Tatiana Romanova, *Stanovlenie Evropeiskogo soyuza kak mezhdunarodnogo aktora. Na primere investitsionnoi deyatelnosti v energeticheskom sotrudnichestve s Rossiei 1994–2001gg* (Sankt-Peterburg: Izdatelstvo Sankt-Peterburgskogo universiteta, 2003).

52 Karmodi, 'Ne truboi edinoi'.

53 President Putin has set the target of increasing the share of nuclear energy in Russia's energy balance 'at least 20 percent during the initial phase': Vladimir Putin, 'Speech at Meeting with the G8 Energy Ministers', 16 March 2006, <http://president.kremlin.ru/eng/speeches/2006/03/16/1302_type82912type82914_103208.shtml>.

the EU. They do not perhaps originate in the energy dialogue, but have a direct relevance for the prospects of energy partnership between the two parties. The logic of interdependence which at the turn of the century might have seemed the only possible way of conceptualizing the relations between Moscow and Brussels in the energy sector, does not really work because of the intrusion of profoundly political questions of political boundaries between the self and other.

Mutual othering between the EU and Russia is a constitutive phenomenon, leading to a situation where the identity of each party becomes crucially dependent on the image of the other as a geopolitical competitor, and potentially an adversary. The positive image of the EU, so common for the Russian discourse of the mid–1990s, is melting into the undifferentiated figure of the west as the eternal antagonist. Conversely, the current pan-European discourse (as well as the western one, generally speaking), is recreating the Cold War descriptions of Russia as an inherently authoritarian state looming large in the backyard of Europe as a relic of its undemocratic and militaristic past. Instead of a 'Europe whole and free' we end up in a situation where Europe is split into two, very much like in the era of the iron curtain. Intensification of the political antagonism leads to a total mobilization of all available resources, and in this scenario energy inevitably ceases to be a field where the logic of mutual compatibility dominates. Indeed, given the significance of energy for Russia's economy and its role in global affairs, this policy sector becomes increasingly politicized and even securitized. Hence, after the events of late 2005–early 2006 energy has once and for all established itself as one of the key themes of Europe's political narratives, where pragmatic logic of economic co-operation no longer applies adequately.

Some authors have suggested that the way out of this dead end must be sought along the lines of the neomedieval model, which envisages a Europe of better empowered regions, multiple identities coexisting rather than engaging in mutual exclusion, and of overlapping political spaces.[54] It is, however, very obvious at this stage that the Northern Dimension, which is usually cited as an example of this approach, has been marginalized in the debate about the future of Europe, and for the most part absorbed by the imperial discourse. The economic potential of the ND is quite substantial, but, as in many other energy-related fields, it is heavily dependent on mutual understanding in the political domain.[55] As already mentioned, the EU bureaucracy has tried to develop the ENP as a solution to the internal/external security paradox, but failed to come up with anything really innovative in comparison to the old logic of conditionality based on the premises of modernization theory. Once again, this is not to argue that modernization theory is entirely wrong, but rather to emphasize how the policy of imposing one particular version of modernity with little sensitivity to local concerns has its limits. It simply does not work in the case of Russia, whose self-image is based on a very firmly established narrative of a sovereign, and in many respects self-sufficient, European great power.

---

54 Joenniemi and Browning, 'Discourses on Centrality and Marginality'.

55 Tero Lausala, 'The Role of Energy in the Northern Dimension', in Lassi Heininen (ed.), *Northern Borders and Security – Dimensions for Regional Cooperation and Interdependence* (Turku: Turku School of Economics and Business Administration, 2002), esp. p. 207.

The Russian decision-makers, too, remain locked in the Westphalian understanding of Europe and unable to grasp the opportunities afforded by the fundamental transformation of European political landscape after the end of the Cold War. Even when the important advantages of deeper co-operation with the EU in all fields including energy are recognized, all too often they are interpreted in accordance with the zero-sum game logic, which makes any 'concession' look as a net damage to Russia's national security. Another effect of the Westphalian approach to Europe is that Russian policy-makers and diplomats prefer to deal bilaterally with the European 'great powers' (Germany, France, Britain) and the EU as a whole, while the northern 'periphery' attracts very little attention. This is obviously one of the reasons why Russia has never been really willing to engage with the ND initiative.[56] An additional explanation here, and in the field of the energy dialogue more specifically, is that the Russian political system remains tuned to maximizing short-term outcomes, while the long-term goals, be it the search for new technologies or doing away with the legacy of Cold War, remain beyond the horizon.

What follows from this analysis is that it is up to the peripheries themselves to mobilize and create alliances in favour of a less centralized and more open Europe. Pertti Joenniemi and Christopher S. Browning argue that Finland should be particularly interested in continuing to support the ND which it helped to create, whilst concomitantly supporting similar approaches to regional co-operation in the European north.[57] Finland has, indeed, been especially keen on promoting cross-border co-operation with Russia, including in the area of the environment, whereas in the energy sector the record is more mixed (see Chapter 6).

There are also some hopes that the regions of Russia's northwest, in particular St. Petersburg, might possess the resources necessary for challenging the monopoly of the federal centre without necessarily questioning the territorial integrity of Russia.[58] Due to their geographical location, the northwestern territories not only serve as an interface between Russia and the EU, but also engage in various kinds of networking with their EU neighbours, such as the Euroregions, the Baltic Sea States Subregional Co-operation, and the Union of Baltic Cities. Various historical narratives – such as the story of Hansa or of St. Petersburg as a 'window to Europe' are employed in order to create a shared regional identity. History, however, does not always play a 'positive' role: on the contrary, conflicting interpretations of the past are a source of continuous tension between Russia and the Baltic States and, to some extent, also Finland. This tension constitutes an important part of national identity construction. But at the same time, by hampering cross-border interaction, the same tension creates tangible problems for the people in the peripheral areas.

56 Joenniemi and Browning, 'Discourses on Centrality and Marginality'.

57 Ibid.

58 Pertti Joenniemi (ed.), *Saint-Petersburg – Russian, European and Beyond* (St. Petersburg: St. Petersburg State University Press, 2001); Viatcheslav Morozov, 'The Discourses of St. Petersburg and the Shaping of a Wider Europe: Territory, Space and Post-Sovereign Politics', *COPRI Working Papers* 13/2002; Anaïs Marin, 'Integration without Joining? Neighbourhood Relations at the Finnish–Russian Border', *DIIS Working Paper* 14/2006, EUBorderConf series <http://www.diis.dk/graphics/Publications/WP2006/WP2006–14.pdf>.

The discrepancy between national and local politics creates potential for political mobilization in the peripheries.

This argument is fully relevant in the context of the energy dialogue, as the Baltic Sea region is a major outlet for Russia's energy exports to the EU, as well as the site of multiple controversies around the distribution of export flows and their safety. Russia is refusing to use Latvian ports for oil shipments citing violations of the rights of the Russian speakers as the reason. Finland is worried about the environmental impact of the intensifying tanker traffic in the Gulf of Finland. Estonia is refusing to buy Russian oil shale to its electricity generation plants because of the Kyoto obligations. This in turn threatens to plunge some of the neighbouring districts of the Leningrad oblast into depression. The Nord Stream pipeline under the Baltic Sea also has clear political implications. These are only the most visible issues where 'politics' intervenes with 'economics' in northern Europe (see also Chapters 5–8).

The existing conflict potential can and must be considered as constraining in some respects, but empowering in others, because the multiplicity of conflicts with their manifold dimensions – local, regional, pan-European – opens up numerous possibilities for creating new alliances on the basis of new identities. Centralization of the Russian political system under President Putin might have deprived the regional authorities of real political sway, but the 'vertical of power' can hardly be expected to address all the local issues, especially those which may be insignificant for the centre but greatly influence the life chances of the local population. Yet this strategy for empowering the local actors, again, requires going far beyond the narrow field of energy into a much wider political landscape.

Chapter 4

Energy Dialogue from Strategic Partnership to the Regional Level of the Northern Dimension

Tatiana Romanova

Introduction

This chapter suggests that EU–Russia energy relations in northern Europe must be seen as springing from interaction between two different levels. On the one hand, there is the level of strategic partnership between the EU and Russia, or Moscow and Brussels, where the agenda of the EU–Russia energy dialogue was originally designed and developed, and a vision set for it. On the other hand, there is the regional level where the two parties in practice interact and implement the energy dialogue. In this paper, the northern European regional level or the Northern Dimension area is selected for a more specific analysis due to its status as a 'pilot region' in the development of the energy dialogue. The major openings and problems of the energy dialogue can well be identified here, but importantly, they are also conjoined by some new experiments and initiatives. The main focus of the chapter will be on multilateral co-operation (for bilateral relations, see Chapters 5–7).

The energy sector co-operation linking the strategic partnership and regional levels has emerged in northern Europe relatively recently. Regional energy co-operation has taken place in northern Europe long before the energy dialogue was launched, and thus the regionally established links are quite strong. However, today, with the emergence of the EU as an all-European actor with a reach into northern Europe, it is the strategic partnership level between the Union and Russia where the guidance, as well as the political and legal framework for regional energy relations is set. In this way, in northern Europe historically established co-operation patterns enmesh with new ones and for their part shape the conduct of the overall energy dialogue.

The European Union announced in 2003 that the Northern Dimension might serve as a testing ground of energy co-operation in Europe, including the energy dialogue, and thus made the ND area of heightened EU interest.[1] The Russian party was more tacit and ambivalent in this instance. Therefore, the questions to be solved in this

1 European Commission, 'On the Development of Energy Policy for the Enlarged European Union: its Neighbours and Partner Countries', communication from the Commission to the Council and the European Parliament, COM(2003) 262 final, Brussels, May 2003.

chapter are: how are the priorities of the energy dialogue linked to regional issues? Does this co-operation in the ND area contradict or reinforce the energy dialogue's agenda, and can it be used to promote further development of the dialogue?

First, I will briefly describe the background conditions for the initiation of the strategic level of the energy dialogue and its institutional set-up, including the main actors and their interests. Second, I will specify the agenda of the dialogue and trace the development of different issues at the strategic level. Subsequently, I will proceed to demonstrate the key energy actors and their interests on the regional level. Finally I will address the question of how strategic issues enter regional level co-operation, and whether the strategic level is reinforced or weakened by co-operation in the north.

The Strategic Level of EU–Russia Energy Dialogue

The year 2000 was a turning point in EU–Russia relations. First, Russia's leadership change from Boris Yeltsin to Vladimir Putin paved the way for a fresh start. Second, in the 2000 Feira European Council meeting, political conditionality and economic co-operation with Russia were decoupled, allowing the intensification of economic links.[2] Third, the Tacis programme was restructured to support initiatives of the biannual EU–Russia summits, and the EU–Russia Co-operation committee and its sub-committees. These measures helped to create a more pragmatic attitude between the parties. Energy moved into the foreground of co-operation, and the energy dialogue developed out of the two parties' wish to enhance their relations by moving from the declaratory stage to practical co-operation.

The first signal from the EU for the initiation of a new kind of energy partnership was given in a meeting between the President of the European Commission Romano Prodi and Russia's Vice-Premier Viktor Khristenko in September 2000. The idea of an energy dialogue was officially presented in the Paris EU–Russia summit in October:

> The European Union and Russia have decided to institute, on a regular basis, an energy dialogue which will enable progress to be made in the definition of an EU–Russia energy partnership and arrangements for it. This will provide an opportunity to raise all the questions of common interest relating to the sector, including the introduction of

---

2 This important aspect was originally discovered during the interviews conducted by the author in the European Commission and the Council of Ministers' Secretariat in 2000–1. The Feira presidency conclusions put the political demands and economic cooperation under different headings, implying that they should be interpreted through these different lenses. Also note that the conclusions say 'the Union stands ready to assist Russia in meeting these aspirations, objectives and commitments, and to develop cooperation with it on the basis of its Common Strategy and the Partnership and Cooperation Agreement. To this end, the European Council invites the Council and the Commission to review the situation in July and to take the necessary decisions about TACIS and other instruments.'; See Santa Maria da Feira European Council, 'Presidency Conclusions', 19 June 2000, no. 200/1/00, <http://www.consilium.europa.eu/cms3_applications/Applications/newsRoom/loadBook.asp?target=2000&bid=76&lang=1&cmsId=347>.

cooperation on energy saving, rationalisation of production and transport infrastructures, European investment possibilities, and relations between producer and consumer countries.[3]

The idea of an energy dialogue needed further development by both parties. They decided to institute a group of two sole interlocutors, and four working groups (energy strategies and balances of co-operation, investments, infrastructure and technology, and energy efficiency). With the development of the energy dialogue's agenda the four working groups were dissolved in 2001. The two sole interlocutors continue their functions. The dialogue was then integrated into the activities of the EU–Russia sub-committee on energy, environment, and nuclear co-operation of the Co-operation Committee, while the two sole interlocutors continue to monitor the realization of the energy dialogue, and periodically produce progress reports for EU–Russia summits.[4]

In 2002, the EU–Russia Energy Technology Centre was established to facilitate communication flows between entrepreneurs in the EU and Russia, and to link strategic issues with practical down-to-earth co-operation. Four new temporary groups (investment, development of energy infrastructure, improvement of trade in energy products, energy saving, and energy efficiency) were formed between 2004–6 with a view to updating the energy dialogue's agenda.

The EU Side

The European Commission's Directorate General on Transport and Energy (DG TREN) handled the practicalities of launching the energy dialogue. Upon its launch DG TREN was busy preparing the 2000 Green Paper on energy security. One of its goals was to gradually bring the question of the security of energy supplies under the competence of the European Communities, at the expense of member-states' sovereignty. The Green Paper was published in November 2000, a month after the EU–Russia summit.[5] It was prepared by the same unit of DG TREN, headed by Christian Cleutinx, which originally came up with the idea of EU–Russia energy dialogue.[6] Thus the dialogue was intended as the first initiative in developing a new external energy policy for the EU.

---

3 'Joint Declaration of the President of the European Council, Mr. J. Chirac, assisted by the Secretary-General of the Council/High Representative for the Common Foreign and Security Policy of the EU, Mr. J. Solana, of the President of the Commission of the European Communities, Mr. R. Prodi, and of the President of the Russian Federation, Mr. V.V. Putin', Paris, 30 October 2000.

4 At the time of writing, six further reports have been approved: in May and November 2002, November 2003, November 2004, October 2005, and November 2006.

5 European Commission, 'Green Paper: Towards a European Strategy for the Security of Energy Supply', COM(2000) 769 final, Brussels, 29 November 2000.

6 Christian Cleutinx continued to coordinate the EU–Russia energy dialogue in the EU Commission even after becoming the director responsible for nuclear energy within the Commission.

The Commission presented a strong case in its 2000 Green Paper. It calculated that the EU's energy demand grew at a rate between 1–2 per cent a year since 1986, while the EU itself was home to a mere 2 per cent of world resources, equivalent to only 20 per cent of EU consumption. By the year 2020, according to the Green Paper, the dependence on external energy supplies of the EU–15 would increase to 90 per cent for oil, 70 per cent for gas, and 100 per cent for coal. The Union's 2004 enlargement would weaken the situation. As for oil, the dependence was set to grow to 94 per cent, and for gas, 90 per cent. In all, the document put forth the often quoted argument that if no measures were taken, by 2020–30, 70 per cent of the Union's energy requirements would be covered by imports.[7]

Concerns for the security of energy supplies were to be addressed through market means. In other words, the Commission's idea was that the means of public administration should be directed at the promotion of market principles in order to guarantee energy companies' access to different markets. The provision of a stable regulatory regime for companies would be the foremost task. The Commission's proposal for Russia was familiar from the ECT (see Chapter 1), and intended to spread the single market beyond the EU's own borders – '[t]he Community *acquis* could become a reference framework for a reform of the energy sector to be implemented in Russia … The EU considers that this reform should be a priority for Russia …'[8] While EU member states in the general sense agreed with this agenda, they found it problematic to cede additional competences to the Commission. Security of energy supplies was a touchy aspect nationally. Therefore, although taking an active part in the discussion on the Green Paper, member states made sure that the Commission did not proceed too far and quickly with the implementation of the EU–Russia energy dialogue. They agreed to appoint Francois Lamoureux, Director General of DG TREN, to the position of EU sole interlocutor in the energy dialogue, but as an administrative, not as a political figure (like a Commissioner). This stressed the technical character of the dialogue. Also, member states only agreed to give the Commission a free hand in the initial design of the energy dialogue's agenda for a short period of six months. The work was done by the four working groups (see above) with representation from both the EU and Russia at the level of officials (including member states' officials), energy experts and business representatives. Member states supervised the activity closely. For example, when one of the working groups had to meet once more after the six months period had expired, the British Government refused to pay the expenses of its own expert, and an official from the British Embassy went to the meeting instead.[9] On the whole, the Commission had to be very cautious in choosing topics for the dialogue. Partly for this reason it decided to concentrate on issues that were either very technical, therefore clearly falling into

7 European Commission, 'Green Paper: Towards a European Strategy for the Security of Energy Supply'.

8 European Commission, 'Communication to the Council and the European Parliament: The Energy Dialogue between the European Union and the Russian Federation between 2000 and 2004', COM(2004) 777 final, Brussels, 13 December 2004.

9 Interview with a European Commission official, Brussels, March 2001.

its own competence, or were quite abstract, long-term, and ambitious, thereby not calling for immediate implementation, like a common EU–Russia energy market.

From the very beginning, the Commission emphasized the involvement of business representatives in the dialogue. In fact, businesses associations like Eurelectric for electricity companies, OGP for oil and gas producers, and Eurogas for gas industry were the Commission's strategic partners. In 2003 a steering group of EU and Russian business representatives was created to advise the two sole interlocutors. Moreover, when primarily on the Commission's initiative the four new working groups were created in 2004–5, they included more business representatives than the 2001 working groups. This affected the discussions significantly.[10] Finally, the working groups reported to the EU–Russian Roundtable of Industrialists in 2005. Its summary of the findings was approved in the conclusions of the EU–Russia Permanent Co-operation Council on Energy.[11] However, in the course of this greater business involvement, the Commission mostly relied on big companies, with evident reflections on the issues it pursued in the dialogue.

In summary, the Commission is the EU's main representative in the energy dialogue, and devises the agenda under close supervision of the member states and in co-operation with representatives of big businesses. Francois Lamoureux was DG TREN's Director General until January 2006 and served as the EU's face in the dialogue, as the Union's sole interlocutor.[12] Difficult internal negotiations between the Commission, member states and businesses finally helped speaking with one voice. The Commission pushed mostly technocratic aspects that were of interest to the big businesses, but alongside these, it included some long-term and abstract goals of a future EU–Russia common energy market, and promoted the incorporation of energy efficiency issues and the Kyoto protocol. The importance of the Kyoto protocol grew within EU–Russia relations, but it never assumed primacy on the agenda because it was beyond the scope of responsibility of DG TREN.

The Russian Side

Russia's Energy Strategy until 2020 identifies several priority axes.[13] The aims include: efficient use of primary resources; upgrading energy technologies for exploration; mining and transit, promotion of market principles (in particular, reforming natural monopolies and making them more transparent in their operations); competition; development of new resources; increase of coal use for electricity generation; and diversification of export markets and a more persistent strategy of integration into the global markets. Russia aims to preserve its position as the main exporter to the EU market during its ongoing liberalization and enlargement. Long-term gas

10 Interview with Christian Cleutinx, Brussels, 28 November 2005.

11 'First Meeting of the Permanent Partnership Council on Energy between the EU and Russia', London, 3 October 2005, 12902/05 (Presse 253).

12 This policy was adopted by the Energy Commissioner Andris Piebalgs in January 2006.

13 The Russian Federation, 'Energeticheskaya strategiya Rossii na period do 2020 goda', Moscow, May 2003.

contracts are crucial here as is the movement to the gas distribution sector. But alongside raw materials, Russia wants to export processed products like electricity and nuclear materials. This is against the interests of industries based in the EU area. Simultaneously as Russia strives for new European investments, it intends to preserve its independence in the regulation of its internal market.

This is conjoined by Russia's reluctance to accept EU proposals for a common regulatory regime based on *acquis communautaire* as a long-term foundation for mutual energy relations. The vision of the Russian side was clearly formulated at the G8 summit in St. Petersburg in July 2006: international norms in the field of energy should take into consideration the rights and interests of both producers and consumers and both sides have to participate in the creation of new norms. Possibilities of investments in exploration can only be provided if there is a reciprocal opening of the energy distribution businesses within the EU.[14]

The Russian side looks at the outset more unified compared to the EU's internal bargaining processes. Russia is mainly represented by the Ministry of Energy, Ministry of Economic Development and Trade, Ministry of Natural Resources and the Ministry of Foreign Affairs. Representatives of two key companies interested in the energy dialogue are included (Gazprom, the gas monopolist, and RAO UES, the electricity company). However, due to the fact that neither the gas nor the electricity sector are reformed, and that the state has majority stakes in both companies, their representatives mostly support the official line. Both public and private sectors are interested in preserving and enlarging the influence of Russia's energy sector in the EU's energy markets, as well as securing Russia's own right to reform the domestic sector at a convenient speed. Energy efficiency and Kyoto goals are low on the Russian agenda, although RAO UES has more interests in this question because it has licensed emissions for sale to European enterprises. This includes some controversies, for example in the case of RAO UES's ambitions to enlarge its export of electricity to the EU. Gazprom felt that it could potentially violate its interests, because the exported electricity would be produced from cheap domestic gas. Hence, Gazprom would get a part of the profit made by RAO UES.[15]

With the growing influence of private actors in the energy dialogue, the roles of RAO UES and Gazprom were also upgraded. At the same time the envisaged liberalization of Russia's gas and electricity sectors made little progress. In fact, the Yukos case as well as Gazprom's acquisition of Sibneft pointed to increasing de-privatization of the sector. This in turn means that the Russian energy sector is unlikely to command an independent voice in the EU–Russia energy dialogue for the foreseeable future. The unity of the Russian position has been further guaranteed by the Russian sole interlocutor Victor Khristenko. At different stages he has occupied the positions of Vice-deputy Prime Minister of the Russian Government and Minister of Energy before taking his present post of Minister of Industry and Energy.

---

14 'Globalnaya energeticheskaya bezopaznost', prinyato 16 iulya 2006 goda v Sankt-Peterburge liderami ctran "Gruphy vosmi".

15 Illustrative examples include the discussion between official representatives of RAO UES and Gazprom that accompanied the construction of the new power plant in Kaliningrad as well as export of electricity generated in the northwestern power plant to Finland.

The Evolution of the Energy Dialogue's Agenda

The activities of the first four EU–Russia working groups resulted in a synthesis report presented in the October 2001 summit.[16] The report confirmed the complementing interests of the two parties, and divided them into short-term goals (legal guarantees for long-term supply, transport infrastructure, enhancement of current legal framework for investments, and energy efficiency) and long-term goals (liberalisation of Russia's gas and electricity market, EU–Russia electricity trade and synchronisation of networks, technological co-operation and implementation of the Kyoto protocol). Gradually the number of items was reduced, and the distinction between short-term and long-term issues virtually disappeared. Discussion focused on five key issues of which the last one remains long-term and more abstract.

Markets and Security of Supplies

This issue was initiated through discussions on long-term contracts for natural gas supply and trade in nuclear materials. Since the 1970s, natural gas supply was based on long-term contracts agreed for a period of 15–20 years in order to divide the risks of costly pipeline construction and resource development between producers and consumers. Many of these contracts contained a clause prohibiting European companies to resell gas outside their home country, so as to prevent Russian gas from competing with itself. These provisions contradicted market principles and were in conflict with the EU's liberalization process in the energy sphere. Therefore, they were raised in EU–Russia talks following the adoption of the directives EC 1998/30 and EC 2003/55 on the liberalization of the natural gas sector in the EU.

The Commission insisted on shortening the duration of the contracts and on leaving the destination clause out, while major EU gas companies were against abrupt action.[17] However, the Commission's threat to fine European companies functioned as a good stimulus for putting additional pressure on Gazprom. The DG Competition initiated the so called ENI/Gazprom case against Gazprom. Finally Russian officials together with Gazprom's representatives had to agree with the EU-induced new rules. In exchange, the Commission agreed to recognize long-term contracts as an essential element for stability in the EU gas market. This arrangement allowed Gazprom to secure a central place in the liberalizing EU market. Gazprom also obtained the right of access to major pipelines in EU member states and to sell

16 'EU–Russian Energy Dialogue: Synthesis Report', presented by Russian Vice-Prime Minister Victor Khristenko and European Commission Director-General François Lamoureux, Brussels/Moscow, September 2001.

17 See e.g. the discussions between Gazprom and the German company Wintershall, on the one hand, and the former Commissioner for Energy and Transport Loyola de Palacio, on the other, Strasbourg; 'Peregovory s Evrokomissies na vysokom urovne', Gazprom, 5 April 2004; 'Gazprom obeschaet povysit nadezhnost postavok gaza v Evropu', Rosbalt 5 April 2004; 'Znachenie Rossii dlya obespechenia energobezopasnosti EC vozrastaet, schitaet rukovoditel SP Wingas', ITAR-TASS, 7 April 2004; 'Ultimatum vysokogo urovnya', Vremya novostei 6 April 2004.

gas directly to consumers, potentially the most profitable segment of the gas market.[18] Finally, Gazprom was granted the right to insert a clause in contracts allowing it to retain its profit margin in the event of gas sales to third countries.

The issue of trade in nuclear materials dates back to 1992, when the Euratom Supply Agency decided to limit the import of enriched fuel to the European Union to 20 per cent. This decision constituted a trade defence measure prompted by the fear of Russia dumping cheap products into the EU market. Since the conclusion of the PCA agreement, Russia has insisted on reaching a mutually acceptable solution on the basis of its article 22. In November 2002, the European Commission presented to the Council a proposal for obtaining a mandate for negotiations with Russia. The discussions finally started in 2004, and in May the Commission forwarded its draft to the Russian party. However, at the end of 2006 the question remained unresolved, largely due to the strength of the nuclear lobby in the EU. The Commission continued to use the issue of nuclear safety as a pretext for limiting access of Russian materials into the Union.

The discussions on the security of supplies eliminated Russia's fears that the EU would limit imports from any source if its share exceeded 30 per cent of the EU's imports. No legal stipulation exists for such a percentage limit. Thus the fear was artificially created by the Russian side. However, security of supplies issues are bound to stay on the agenda as a result of the 2005–6 conflict between Russia and Ukraine on gas prices for Ukraine and transit of gas through its territory (see below).

Transport Infrastructure

Transport infrastructure is in the interest of both parties because it is an essential element of exploiting Russia's resources. However, with the accession of new member states to the EU, the issue has become more controversial due to the fact that many new member states that are transit countries consider their transit role as one of the key factors in their energy security (for the Baltic states, see Chapter 7).

EU market principles dictate that companies are responsible for the construction of new infrastructure. Concomitantly the principle of security of energy supplies requires that consumers are guaranteed a stable supply. Provision of infrastructure hence becomes essential. Thus the parties of the energy dialogue decided to accord the status of a project of common interest for the completion of the Yamal–Europe gas pipeline network through Belarus and Poland, for the Nord Stream gas pipeline from Russia to Germany, for the connection of the Druzhba oil transmission system through Belarus and Ukraine with the Adria network, and for the interconnection of the parties' electricity networks, which would ensure non-discriminatory transit of energy products and increased supplies to the EU and prospective member states.

---

18 The remaining problem for Gazprom is the unwillingness of European consumers to change their suppliers; see e.g. European Commission, 'Report from the Commission on the Implementation of the Gas and Electricity Internal Market', COM(2004) 863 final, Brussels, 5 January 2005.

The completion of the second line of the Yamal–Europe pipeline and the construction of the Nord Stream pipeline contradict each other in the short-term. The Nord Stream project also attracts a lot of criticism in the Baltic states (see below; Chapters 5 and 7). Interconnection and synchronisation of electricity systems is in the interest of the Russian RAO UES, whose excessive capacities and low environmental costs afford producing low cost energy and selling it for a high price in the EU market.[19]

The status of a project of common interest signifies political and administrative support from the partners. These projects are supported through the organization of round tables between government officials, financial institutions and energy companies in order to promote dialogue and facilitate the search for partners when large-scale funding is needed. Further support measures include the EU's internal policy of Trans-European Networks (TEN). The revision of priorities in 2003 and the definition of priority axes took into full account the selected projects of common interest.[20] The budget line of the TEN allows for financing technical and feasibility studies. Projects are eligible for low interest credits from the European Investment Bank (EIB). Furthermore, the EU envisages investment, mainly on its own side, along the priority axes of the TEN-Energy until 2013.[21] The new working group on infrastructure that was set up in 2005 suggested examining possible new projects and issues of equal access to transport grids.

Environmental aspects played a visible role in the development of infrastructure projects. The Erika and Prestige oil tankers' accidents in 1999 and 2002 made the EU aware of the dangers of oil transportation by sea. The Commission firmly insisted on tougher controls, and eventual banned single-hull tankers in its territorial waters.[22]

The Commission tried to promote ground transportation as a further measure for guaranteeing its physical security. For Russia, the issue of ground transportation equates to dependence on transit countries like Ukraine and Belarus, plus the related transit charges, and illegal use of transit oil and natural gas. The conflict

---

19 Currently the Russian United Power System (UPS/IPS) is not synchronized with the EU's Union for the Co-ordination of Transmission of Electricity (UCTE) and its two satellites (NORD-EL and CENTREL). Synchronization would allow for an unhampered flow of electricity. This issue was strongly lobbied by Russian producers and political agreement was finally reached in October–November 2003. A working group was formed to prepare a feasibility study by 2007. In December 2003 RAO UES prepared a first draft of the concept, defining options for synchronization of Russian electricity networks with UCTE (and NORD-EL, CENTREL). But the energy crisis in Russia during the cold winter of 2005–6 severely undermined the credibility of increased electricity export offers from Russia to the EU.

20 See European Parliament, Council of Ministers, 'Decision Laying down a Series of Guidelines for Trans-European Energy Networks and Repealing Decision', no. 1254/96/EC; no. 1229/2003/EC, Brussels, 26 June 2003.

21 See European Commission, 'Envisaged Investment along the Priority Axes of Trans-European Energy Networks (TEN-E)', Brussels, 25 July 2003, <http://europa.eu.int/comm/energy/ten-e/en/index.html>.

22 Also the International Maritime Organisation was involved; see 'Maritime Transport of the European Union', <http://www.europa.eu.int/comm/transport/maritime/safety/2000_erika_en.htm>.

potential on ground transit issues was vividly demonstrated in several episodes, most pronouncedly in Russo–Belarusian energy relations in February 2004 and January 2007, and in Russian–Ukrainian relations in early 2006 (see below). In these circumstances, the construction of the Nord Stream pipeline will help bypass transit countries and enhance security of supplies, but it will proceed at the expense of the environmental and other concerns of Poland and the Baltic States (see Chapters 5, 6 and 7).

Finally, the European Union insisted on using the space navigation system Galileo for the maintenance of gas and oil pipelines, and high-tension electricity. The need for increased monitoring is a result of the fact that most of Russian transportation routes and pipelines are over thirty years old.

Enhancement of Legal Framework, Investment Climate and Transfer of Technologies

The efforts to improve the legal framework for intensified EU–Russia energy relations were hit hard by the great problems with the attempted adoption of a production sharing agreement procedure in Russia. The reform of Russia's tax code and successful lobbying by Russian oil companies resulted in notable changes into the production sharing agreement laws, restricting foreign companies' access to Russian resources. An exception was made for exploration in Russia's Arctic shelf.[23] This provoked a new issue in the report of the new working group on investment: stimulating growth of confidence in Russia's tax system. The discussion continued on facilitating registration and license approval in Russia in order to encourage European companies to invest in Russia's exploration and production facilities. The possibility of a one-stop shop to fulfil all the formalities was discussed, but with no results.

In the field of investment climate, the possibility of a non-commercial risk guarantee fund was turned down by the European Commission in 2004 because it was seen as too costly, and legally complicated in requiring ratification by all EU member states and Russia, plus that it indirectly approved the poor functioning of Russia's judicial system. The Commission suggested a different scheme in 2004, an Energy Desk with the participation of relevant development banks and the European Investment Fund for the initiation of finance for long-term energy projects.[24]

The problems in improving the legal framework and investment climate are striking as both parties are in principle in favour of investment support initiatives. But they operate with very different arguments when it comes to more specific issues under this topic. The European Commission seeks to help EU-based companies, which in turn seek to participate in the exploration of natural resources to guarantee supply to their clients. Here the Commission also wants to promote its fundamental

23 See e.g. <http://www.concession.ru/>; see also Russian Federation, 'Russia's Tax Code and Federal Law on Production Sharing Agreement No. 225FZ' of 30 December 1995 (with changes of 7 January 1999, 6 June 2003, 29 June 2004).

24 European Commission, 'Communication to the Council and the European Parliament: The Energy Dialogue between the European Union and the Russian Federation between 2000 and 2004'.

goal of a pan-European energy market (see Chapter 1). Russia strives to attract investments in specific areas: enhancement of production in existing sites, upgrading of oil refineries, building new and upgrading old power plants, and developing and upgrading the energy transportation infrastructure.[25] Thus, Russia's targets include sites with very low rate of return, and the construction and upgrading of objects that would help Russia to transfer from mere raw materials exports to exporting processed energy goods.

Apart from the legal and financial side, the discussions on investment promotion proceed in the field of technology transfers. The initiatives include use of depleted fields and related technologies; improvement of transportation and production technologies to minimize environmental impact and waste of energy resources; low-emission exploitation of coal; and enhancement of nuclear energy's safety. All these initiatives enjoy mutual support, but have stalled due to the poor investment climate.

The most visible result in this sphere is the establishment of the Russian–European Energy Technology Centre in Moscow in November 2002. Its objective is to co-ordinate all energy projects with EU participation in Russia, and to serve as a link between political decision-making and small and medium business, research institutes and regional energy companies. At the same time, the Centre is part of the Organization for Promotion of Energy Technologies in the European Union, thereby guaranteeing Russia's access to new technologies and practices within the EU. The work of the Centre is structured along three work packages: hydrocarbons, electricity and coal, and energy efficiency and renewables. The Centre also serves as a discussion point on energy standards, technical norms and renewable technologies. Its establishment is one of the few things done within the framework of the energy dialogue for the benefit of small and medium business.

Energy Efficiency and Kyoto Protocol

Energy efficiency did not initially play a prominent role in EU–Russia energy discussions. Yet, concrete steps include Tacis energy efficiency projects in Astrakhan, Archangelsk and Kaliningrad, that demonstrate the potential for energy saving and energy efficiency, and encourage energy savings and spreading of European technologies in Russia. Another project pertains to renewable energy and regeneration of small hydropower stations.

The low profile of energy efficiency in the dialogue started to change after the appointment of the new European Commission in 2004. The reference to the EU's Green Paper on energy efficiency in the sixth progress report on the energy dialogue was noteworthy in this respect.[26] The new working group on energy efficiency suggested exploring ways to increase the capacity of the EU–Russia Energy Technology Centre for the promotion of innovate and state of the art technology for

---

25 EU–Russian Energy Dialogue, 'Sixth Progress Report', presented by Russian Minister of Industry and Energy Victor Khristenko and European Commission Director-General Francois Lamoureux, Moscow/ Brussels, October 2005.

26 Ibid.

energy efficiency projects. Examining legislation and incentives to stimulate greater energy saving and to refinance investments into housing were proposed as well.

Co-operation in the field of climate change is now an essential aspect of the energy dialogue. The dialogue was used to persuade Russia to ratify the Kyoto Protocol – after all, one of the key priorities of the EU is to be a global leader in this sphere. The enforcement of the protocol depended on Russia's ratification after the US had turned it down earlier. Russia's ratification in February 2005 was a compromise on the part of Moscow and a demonstration of its wish of belonging to the 'right' reference group. The internal discussion on protocol's ratification was quite heated and most of the specialists in the field of climate change, as well as economists were against it.[27] First, obligations under the Kyoto Protocol limit economic growth. Second, the US was supposed to be the main consumer of Russian emission quotas, and its refusal to participate reduced Russia's market drastically. The EU gives preference to its member states and candidate countries in emissions trade. Hence, economic gains from emissions trade are limited for Russia. Finally, a mechanism to clarify property rights on emissions and trade does not exist in Russia; it is thus unclear who would benefit from the protocol's ratification. But ratification also had its supporters.[28] At the company front, RAO UES had already some emissions licensed within the Kyoto implementation mechanisms. Immediately following the ratification, RAO UES proposed two projects for upgrading its power and heat plants in exchange for its emissions quota. The first contracts were concluded between RAO UES' subsidiaries (Orenburgenergo and Khabarovskenergo) and the Danish Environmental Protection Agency in 2005.[29] There is a huge potential for further co-operation in this sphere provided the demand for emissions rights is high.

---

27 See A. Volsky, 'baz ucheta strategicheskih interesov Rossii nelzya dalshe vesti mezhdunarodnyi peregovornyi process po Kiotskomy protokolu', Agentstvo biznes novostei, 19 February 2004; 'V Gosdume proidut parlamentskie slushaniya o Kiotskom protokole', *Kommersant*, 15 April 2004; 'Vypolnenie Kiotskogo protokola v pervoe vremya ne privedet k znachitelnomu snizheniu vybrosov parnikovyh gazov', Rosgidromet, RIA Novosti, 9 September 2003; 'Illarionov schitaet nepodtverzhdennoi teoriu o reshauschem vliyanii uglekislogo gaza na globalnoe poteplenie', RIA Novosti, 1 October 2003; 'Rossiya ne speshit zanosit sebya v protokol', *Kommersant*, 10 September 2003; 'Rossiya pognalas za "parnikovoi khimeroi"', RIA RBK, 1 October 2004; 'Rossiya poka ne gotova ratifitsirovat Kiotsky protokol, zayavil Putin', Izvestia, 29 September 2003; 'Rossiya ne stavit krest na Kiotskom protokole', <http://www.strana.ru> (6 December 2003).

28 See 'Vybrosy uglekislogo gaza v Rossii sokratilis s 1990 goda na 32 %', RIA Novosti, 25 September 2003; 'Kioto: dengi v obmen na chistotu', Vremya novostei, 21 June 2004; Alexander Kosarikov, 'Kioto otkryvaet biznes', *Neft i kapital*, 2/2003; 'Perspektivy uchastiya Rossii v globalnyh rynochnyh mehanizmah Kiotskogo protokolai privlecheniya investitsii v energoeffektivnye proekty', *Izvestia*, 19 February 2004; 'Prodavtsy vozduha: Rossiya gotova ratifitsirovat Kiotskii protokol', Novye izvestia, 10 September 2003; 'Ratifikatsiya Kiotskogo protokola mozhet prinesti Rossii polzu – minprirody', RIA Novosti, 20 November 2003.

29 The first project presupposes the installation of two 10 MW combined-cycle gas turbines at the Mednogorskaya power plant. As a result of the second project, the Amurskaya power plant will shift from coal use to natural gas and increase the efficiency of pressure.

Establishment of a Pan-European Energy Market

The creation of a pan-European energy market is the only long-term idea with a more abstract nature which the European Commission has pursued within the context of the energy dialogue. It is designed to be implemented by technocratic actions of the European Commission and the relevant authorities in Russia. The Commission initiated discussion on the liberalization of Russia's gas and electricity market, including unbundling of exploration (production), transit and distribution; third-party access; suppliers' competition in creating a market for consumers; and ensuring the right to choose a supplier. This would introduce transparency into Russia's energy sector and serve as an important ingredient of a pan-European market.

In early 2003, the reform of Russian's electricity sector was agreed. The introduction of concrete measures was started in November 2003 with the free trade of small amounts of electricity. Wholesale generating companies were then established. The EU was an active supporter of the reform and made a symbolic gesture by inviting Russia to participate in the Florence Forum of electricity regulators in September 2004. Through this forum the EU tried to promote legal harmonization within the European Union by encouraging companies to follow common rules. However, in 2005 the reform of Russia's electricity sector stalled. Interestingly, the reform, including liberalization and adoption of EU safety standards, was presented as a condition for synchronizing EU and Russian electricity networks. Russia also had to achieve a degree of market openness comparable to that of the EU. This strategy was in a stark contrast to the EU's search for an increased supply of gas.

The reform of Russia's gas sector was the most conflict-inducing issue in Russia and within the wider context. Gazprom was against a radical separation of transportation business from the parent company. The Russian Ministry of Trade and Economic Development supported a liberal point of view on reforming Gazprom to resemble the energy sector in the EU. The reform of the gas sector was postponed for quite some time because the state did not have a majority stake in Gazprom. This question was successfully solved in 2005, and some reforms took place in 2006. They were, however, limited to the free float of Gazprom's shares, increase of prices and small changes in the relations between Gazprom and independent suppliers.

Liberalization of Russia's gas sector was very prominent in the discussions on Russia's WTO accession in 2004. In this framework the DG Trade insisted on raising Russia's domestic gas prices and on the liberalization of gas transportation. These demands within the WTO can be viewed as the EU's effort to extend the sphere of application of the *acquis communautaire* through an international body in order to enforce commercial diversification both of companies supplying natural gas to the EU market and of means of transportation. This would not only lay a corner stone for a future pan-European gas market, but also reallocate substantial gas profits from Russia to European companies.[30] However, the EU failed to push through all its

30 William Tompson, 'The EC's Negotiating Position with Respect to the Russian Energy Sector in the Context of Russia's WTO Accession Bid: Prospects for the Russian Federation Project' (London: The Royal Institute for International Affairs, January 2004); European Commission, 'Pascal Lamy in Russia to discuss WTO accession', Brussels, 15 October 2003;

demands. Gazprom was allowed to keep its export monopoly and the Commission gave up its insistence on liberalizing gas transit. At the same time Russia agreed to gradually raise internal gas prices to 50 US dollars per 1000 cubic meters by 2010, and to withhold raising export duties on natural gas. These questions resurfaced in the G8 discussions on energy security, in the discussions on the ECT as well as on the new agreement between Russia and the European Union.

Liberalizing the electricity and natural gas sector is the centrepiece of reforms to the EU's energy sector. Consequently, these issues also became the legal backbone to the formation of a pan-European market. However, to be effective, legal approximation has to be all-inclusive, and consequently, liberalization on the Russian side will from the EU point of view need to be complemented with the adoption of EU environmental norms and nuclear safety standards, in order to equate systems of regulation so that a pan-European market would function properly. The reason for these demands is that expenditures on environmental regulation and safety constitute approximately 70 per cent of the final electricity price in the EU. In this light, the concerns of EU based electricity producers of Russian electricity being dumped into the EU market become more understandable. The Russian position on legal harmonization is far more cautious. Khristenko stated that in the present situation, full harmonization of economic and legal systems is a long-term issue, and co-operation in individual sectors has to be emphasized.[31] But a new harmonization project of the EU and Russian energy policies was initiated in autumn 2005 and was expected to play 'an important facilitating role' in the creation of a pan-European market.[32]

* * * *

Summing up the discussion at the strategic level, the two parties have a mutual interest in the dialogue, but each pursues its own agenda. For differing reasons both Russia and the EU prefer to reduce the energy dialogue to technical questions. The European Commission tries to guarantee security of energy supplies to the whole Union based on the principles fixed in the Green Paper on energy security from 2000. The Commission promotes enhancement of the legal framework and investment climate in Russia, and the development of transport infrastructure as technical means

---

'Putin Warns EU on WTO Arm-Twisting', *St. Petersburg Times*, 10 October 2003; 'Russia toughens stance on energy prices', *Financial Times*, 16 October 2003; Vladimir Putin, 'Rossiya ne soglasitsya s trebovaniem Evrosouza nemedlenno pereiti ne mirovye tseny v energetike', *Neftegazovaya vertical*, 9 October 2003; 'Evrokomisiya poluchila chetkii signal ot rossiiskoi storony o tom, chto exportnaya monopoliya "Gazproma" sohranitsya – Lami', RIA Novosti, 16 October 2003; 'Pascal Lami zalomil vtridoroga. Zapad predlagaet Rossii kupit chlenstvo v VTO', *Kommersant*, 17 October 2003; 'Podpisanie protokola mezhdu Rossiea i Evrosouzom uskorit process vstupleniya Rossii v VTO – P. Lami', Prime-TASS, 21 May 2004; 'Putin – v gazovoi sfere Evropa budet imet otnoscheniya s rossiiskim gosudarstvom', RIA Novosti, 9 October 2003.

31 Victor Khristenko, 'Doklad V.B. Khristenko na Shestom Obschem Sobranii Kruglogo Stola Promyshlennikov Rossii i Evropeiskogo Souza', The Hague, 10 November 2004.

32 'EU–Russian Energy Dialogue: Sixth Progress Report'.

for creating a pan-European energy market. Gradually energy efficiency and the Kyoto protocol have increased their importance within the energy dialogue.

For long, Russia preferred to solve current problems without considering the strategic future of the EU–Russia energy dialogue. However, recently it started to defend the interests of its energy companies more vigorously. Russia aims at guaranteeing the security of demand for its energy products, to gain stakes in the rapidly changing EU market and preserve its freedom of action both internally and externally. Russia skilfully avoided EU-induced liberalization of its energy sector while pushing for the use of EU funds to finance infrastructure and exploration projects. Discussions on the synchronization of electricity networks were for the most part postponed, and questions on trade in nuclear materials stalled.

The thus traced strategic level policy of the European Commission was until recently more coherent and its alliance with business quite successful. The Russian policy, despite Russia being a far more homogeneous party, only recently emerged in a more coherent form. Until the 2006 G8 summit in St. Petersburg, Russia only managed to mobilize when its interests were in danger: in the case of long-term gas supply contracts, the Nord Stream pipeline and the avoidance of transit states, and avoidance of large-scale legal approximation. As a result, agenda-setting was dominated by the EU's needs and the vision of the Commission, much more so than by Russia's.

The Northern European Regional Level of the Energy Dialogue

The regional level is where the strategic level discussions are transformed into down-to-earth co-operation. The Northern Dimension area is the primary region for the practical implementation of the energy dialogue due to several reasons.

First, two different markets, the liberalizing EU market and the still largely unreformed Russian energy market come into direct contact here due to their geographical interface in northern Europe. Second, the area is a source of natural resources, primarily oil and natural gas, by comprising two key European producers, Russia and Norway (Table 4.1). Third, the area is a transit region for the resources on their way to the bigger markets. The region thus benefits from exploration and trade in energy products, as well as from transit business.

As said, energy co-operation has taken place in northern Europe long before the energy dialogue was launched. As a legacy of the Soviet Union, the Baltic countries and Russia share a single power system as well as a centralized system of gas supply, and partly also oil supply. Since the 1970s, the Soviet Union, and later Russia, has supplied its neighbours gas, oil and electricity; in this way firm links prevail between the Soviet Union/Russia, and Baltic States, Finland, Germany and Poland. Moreover, transit business in the region has stimulated multilateral co-operation. In 1974, the first Helsinki Commission (HELCOM) convention was signed on the protection of the Baltic Sea against all forms of pollution, including oil spills.

In the 1990s, new forms of regional co-operation were set up. They were not specifically energy-oriented, but comprised successful energy co-operation. Energy also came to play a prominent role in the flagship project of the Northern

Dimension. In 1998, the European Commission identified potential spheres of co-operation: 'the long term potential for the exploitation of oil, gas and non-energy raw materials is huge, but will require substantial improvements in energy and transport infrastructure'.[33] Further suggestions were made in 1999.[34]

The first ND Action Plan in 2000 referred to the PCA agreement, which itself was very brief on energy, outlining the principles of market economy coupled with the ECT as the basis of co-operation.[35] Thus, existing priorities were reiterated in the ND documents, with particular emphasis on the issues of reliable exploitation and deliveries of energy resources, energy saving and nuclear safety. But the 2002 progress report on the ND described only limited success in the energy field.[36] The second ND Action Plan put energy co-operation into the wider framework of the EU–Russia Common Economic Space by emphasizing trade and investment. The co-operation under the energy dialogue was to be a testing ground for the CES. As a result, the energy related goals in the second ND Action Plan were much more precise than those in the first plan and correlated closely with the wider EU–Russia energy dialogue. The goals included the development of infrastructure in order to improve security of energy supply; integration of energy markets, and the strengthening of EU–Russia and Norway–Russia energy dialogues, as well as their integration with Baltic Sea regional energy co-operation; sustainability, the promotion of energy efficiency and saving, and possibilities for testing Kyoto flexible mechanisms; improving investment conditions in Russia's energy sector in order to upgrade the infrastructure, promotion of energy efficient and environmentally friendly technologies, and enhancement of energy conservation; monitoring and the protection of the environment from nuclear safety incidents and radiation.[37]

The follow-up to the second Action Plan, which expired at the end of 2006, also contained energy as its natural key part, and this focus was supported by the proposal to establish a special transport partnership fund and re-orientate the existing Northern Dimension Environmental Partnership (NDEP) fund to deal more effectively with the side-effects of energy production and transportation in the region.

---

33 European Commission, 'A Northern Dimension for the Policies of the Union', Communication from the Commission, COM(1998) 589 final, Brussels, 25 November 1998.

34 European Commission, 'Strengthening the Northern Dimension of European Energy Policy', Communication from the Commission, COM(1999) 548 final, Brussels, 8 November 1999.

35 European Council, 'Action Plan for the Northern Dimension with External and Cross-Border Policies of the European Union 2000–2003', Feira, 19–20 June 2000; 'Agreement on Partnership and Cooperation establishing a partnership between the European Communities and their Member States, of the one part, and the Russian Federation, of the other part', Corfu, 24 June 1994; esp. articles 65§1 and 105.

36 European Commission, '2002 Annual Progress Report on the Implementation of the Northern Dimension Action Plan', Commission Staff Working Document, SEC(2002) 1296, Brussels, 26 November 2002.

37 European Council, 'The Second Northern Dimension Action Plan, 2004–2006', Brussels, 16–17 October 2003.

Table 4.1 Energy resources in the northern European region, June 2005

Countries of the region	Oil reserves (trillion tonnes)	Oil reserves (share of total, %)	Natural Gas reserves (trillion cubic meters)	Natural Gas reserves (share of total, %)	Coal deposits (million tonnes)	Coal deposits (share of total, %)
Denmark	0.2	0.1	0.08	0.1	–	–
Estonia	–	–	–	–	–	–
Finland	–	–	–	–	–	–
Germany	–	–	0.32	0.2	66000	6.7
Iceland	–	–	–	–	–	–
Latvia	–	–	–	–	–	–
Lithuania	–	–	–	–	–	–
Norway	1.4	1.0	2.19	1.4	–	–
Poland	–	–	0.17	0.1	22160	2.3
Russia	8.2	5.7	47.57	30.5	157010	15.9
Sweden	–	–	–	–	–	–

Source: 'BP Statistical Review of World Energy', June 2005 <http://www.bp.com>.

Regional Level Actors and their Interests

The first observation is that all public actors that are active at the strategic level are present at the regional level; however, the intensity of their activity varies. The European Commission is hardly active at the regional level. Its main concern is to make sure that the regional agenda does not contradict the one promoted at the strategic level. EU member states are much more active and often act bilaterally rather than multilaterally, the most notorious bilateral case being the German–Russian Nord Stream pipeline (see below; Chapters 5 and 7).

Russia is mainly represented in northern European energy co-operation through its Federal Government. This is striking, as northern European co-operation only impacts northwestern regions of Russia. The Federal Government's involvement must be seen as an attempt to strengthen its authority in Russia, and to keep regional governments at bay in their cross-border relations. The regional governments in northwest Russia accept this centralizing tendency but with some variations (see also Chapter 8). It is noteworthy for example how in 2004, Valentina Matvienko, Governor of St. Petersburg, stressed in an interview that the ND was intended to develop European co-operation in general and to create conditions for the integration of Russia and the EU – although the initial question asked was about the meaning of this co-operation for St. Petersburg, and the ideas of the city regarding it.[38]

The second point of note is that the Baltic Sea Region Energy Co-operation (BASREC) is the key structure in regional intergovernmental energy co-operation. The BASREC was set up following two conferences of the energy ministers of the

38 'Interview s Gubernatorom Sankt-Peterburga Valentinoi Matvienko', BALTINFO, April–May 2004.

Baltic Sea region and the European Commission. As a rule, the Commission is a much more passive participant than the representatives of the eleven countries of the region to whom the discussed issues are more concrete. The first conference in Stavanger in 1998 defined as the goal of energy co-operation the creation of effective, economically and environmentally sound, and more integrated, energy systems in the region. The second conference took place in October 1999 in Helsinki. It formally set up the BASREC, and established a steering group of senior energy officials (GSEO).

BASREC adopted its first working plan for 1999–2002. The participants of BASREC, represented by energy ministers of the region and the European Commission, identified regional electricity and natural gas markets, climate change, energy efficiency and promotion of renewable sources of energy as the main lines of action.[39] The second working plan for 2003–5 further sharpened these goals by suggesting specific targets, and establishing working groups.

BASREC positioned itself as the central organization for the implementation of the energy parts of the ND action plan in an attempt to assume a central role in energy-related activities in the region. BASREC's own working plan was far more concrete and detailed than the energy part of the ND Action Plan.[40] BASREC also tried to tune its practical activity with the points of discussion within the EU–Russia energy dialogue. The organization strongly emphasised creating conditions for co-operation of small and medium enterprises, and enabling co-operation on energy efficiency and climate change. This reflected the influence of Nordic values in BASREC.

The third observation relates to the multiple private actors operating in the northern European region. These are mainly energy companies with interests in their home and neighbouring countries, including Lukoil, Gazprom, RAO UES, Finland's Fortum, German E. ON Ruhrgas and Wintershall, Swedish Vattenfall, Norwegian Statoil and Norsk Hydro. Companies naturally pursue their own agenda of seeking growth and diversification of their business, but their input became essential for some projects like the development of transport infrastructure, exploitation of some resource deposits and realization of the Kyoto joint implementation projects.

Companies also participated actively in debates on the formation of a regional electricity and natural gas market. Two fora were formed. Baltic Gas is an association of 15 gas companies to promote their business in the Baltic region and the development of an integrated natural gas, electricity and CO_2 market. It also works with developing gas transportation infrastructure, introducing gas to new markets and guaranteeing the security of supply through diversification and storage construction. Baltrel was founded by the electricity supply industry in 1998, and includes 16 electricity companies from Belarus, Denmark, Estonia, Latvia, Lithuania, Norway, Poland, Russia and Sweden. Its overall purpose is to work for 'increased co-operation in the

39 'Report by the Group of Senior Energy Officials (GSEO) for the Meeting of the Energy Ministers of the CBSS Member Countries and the European Commission', Vilnius, 20 November 2002.

40 See 'BASREC GSEO Work Plan 2003–2005', <http://www.cbss.st/basrec> (2 September 2003).

electricity industry and the creation of an open and integrated electricity market in the region'.[41] Initially Baltrel concentrated on the completion of the so called Baltic Ring – connecting all electricity networks into a single system which would create the necessary infrastructure for a common regional market. With the adoption of the EU's directive on full liberalization of electricity, ratification of the Kyoto protocol, and development of relevant mechanisms within the EU, Baltrel's activities have shifted to a more general level: market development, enlarging transmission grids and implementation of the Kyoto protocol without contradicting market principles. In this way, a division of responsibilities has developed between energy policy actors in northern Europe.

How Do Strategic Issues Come into Play at the Regional Level?

When thinking of the interrelationships between the strategic and regional levels in EU–Russia energy politics, one must keep in mind how decisions at the strategic level impact regional actors through the division of competences. Even in the absence of full EU competence in energy policy, the northern EU member states have ceded the majority of relevant competences to Brussels, and cannot make bilateral agreements for example on long-term gas contracts and on the supply of nuclear material. And no measure they adopt can contradict EU regulation. But Russia's northwestern regions are even more dependent on the central government in energy issues. Moscow firmly controls the political aspects of regional energy co-operation, as well as aspects of legal harmonization, and discusses them directly with Brussels, or when suitable, with capitals of big EU member states. Consequently, there are limits to what degree the north European region can serve as an alternative arena for EU–Russia energy relations.

In this situation, interrelationships between the strategic and regional levels can be divided into cooperative and conflicting patterns. The cooperative patterns portray northern Europe serving as the most active testing ground for new initiatives where abstract framework decisions are made more concrete, and, from where new ideas can be drawn into the energy dialogue. The conflicting patterns connote northern European interaction going against the logic and provisions of the energy dialogue set at the strategic level.

Co-operative Patterns

(i) *ND energy co-operation as a testing ground for the energy dialogue: development of transport infrastructure.* This is the primary example of northern European co-operation. The countries of the region have a tradition of co-operation driven by regional interdependencies, and thus became pioneers in implementing new regulatory provisions agreed by Russia and the EU. In 2001, Baltic Gas and Baltrel jointly released a study on the development of infrastructure, which called 'to identify ways to improve conditions for a balanced development of the gas and electricity

41 'Baltrel Annual Report 2004', <http://www.baltrel.org>.

infrastructure in the region'.[42] This study provided a further basis and more specific details for developing the trans-European networks. BASREC's working group on natural gas, for its part, mainly concentrates on issues such as monitoring demand and supply in the region to timely propose the development of new gas infrastructure. The group also elaborates the operational aspects of gas infrastructure with an eye on improving the management of networks to increase reliability of gas supply.

The most prominent project in the region is the Nord Stream gas pipeline, pushed through close co-operation between Russian and German public authorities, with the preliminary agreement concluded between Gazprom, E.ON Ruhrgas and BASF in September 2005 after long negotiations and calculations of economic viability. Other projects include the completion of the first line of the Yamal–Europe gas pipeline by EuroPolGas, a joint venture of Gazprom and PGNiG, and an extension of the oil transportation network from Russian fields into its shores in the Baltic Sea.

Activities of the Baltic Gas and Baltrel were pivotal in the preparatory stage, whereas specific companies are responsible for the realization of the actual pipelines and electricity links. Overall the priorities set at the strategic level are reflected in how in questions pertaining to gas, infrastructure issues dominate the discussion, and in how in the electricity sector, regulation is the main subject of discussions. In these issues, the regional level becomes a platform for linking two different legal systems with each other and for testing the practicalities of implementation.

(ii) *ND energy co-operation as an effort to strengthen the energy dialogue: energy efficiency and technological co-operation, enhancement of legal framework and establishment of a pan-European energy market.* In technological co-operation BASREC is the main actor, particularly in areas where its working groups on energy efficiency, and on climate change and bioenergy operate. Energy efficiency is widely seen as connoting a lot of potential within the northern European region. The ND states vary widely in their energy intensity, i.e., the amount of energy used compared to GDP produced (see Figure 4.1).

According to the European Commission, the potential for energy saving in Russia's Kaliningrad region is about 35–40 per cent of the consumption.[43] Although the reasons can be partly found in the region's small size and the structure of its economy, it is clear that western experience in energy efficiency and energy saving can be exploited to develop more environment friendly modes of production in Russia. Two out of the three locations where Tacis grants were provided for energy efficiency projects are located in the ND area (Archangelsk and Kaliningrad). The 2003–5 BASREC working programme contained very precise activities: communicating common policies on energy efficiency, spreading of combusted heat power and district heating technologies, elaboration of an energy efficiency handbook for residential buildings, harmonization of standards in the field of bioenergy, and joint implementation of bioenergy projects.

---

42 'Gas and Electricity in the Baltic Sea Region', Baltic Gas, Baltrel and the European Commission, October 2001.

43 European Commission, 'The Energy Dialogue between the European Union and the Russian Federation between 2000 and 2004'.

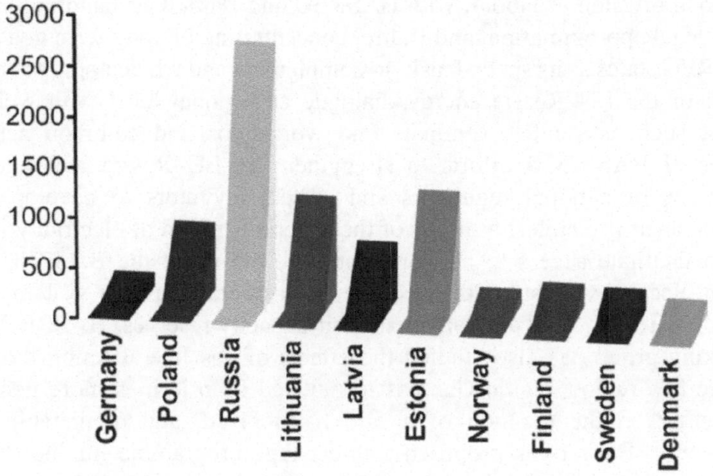

Figure 4.1 Energy intensity among northern European states (tonnes per million dollars of GDP)

Source: European Commission, 'A Northern Dimension for the Policies of the Union', communication from the Commission, Brussels, 25 November 1998, COM (1998) 589 final.

BASREC has set the provision of investments for relevant projects as one of its key tasks. The Nordic countries as well offer different mechanisms to finance small energy efficiency measures in the region. A conference organized in 2004 in St. Petersburg to exchange experience on financing energy efficiency and climate change projects was particularly successful.[44] The role and experience of financial institutions such as the EBRD, Nordic Investment Bank (NIB), World Bank, the NDEP fund, and Nordic Environmental Financing Corporation (working under the guidance of BASREC) is central in providing smaller grants, initially in Poland, followed by the three Baltic countries, and finally in Russia. International financial corporations co-operate on small-scale investments with national financing institutions, initially in the Baltic countries and Poland. In 2004 they announced their preparedness to enter Russia.

Technology transfers are often initiated by companies. The most obvious example is the co-operation between Gazprom and the Norwegian companies Norsk Hydro and Statoil on technologies for the exploration and development of oil and gas fields in the Arctic region. However, unlike projects sponsored by BASREC, technology transfers initiated by big companies are project-specific and are very often used as an argument for assuming a stake in the development of new fields (see Chapter 6).

44 See materials of the conference, 'Northern Dimension in the BASREC Perspective', St. Petersburg, 16–17 March, <http://www.basrec.org/>.

BASREC, Baltic gas and Baltrel play a substantial role in enhancing the legal framework for investments and in developing a pan-European energy market. They have a good division of labour, with BASREC and Baltic Gas handling immediate issues of legal approximation, and Baltrel concentrating on long-term goals.

BASREC states as its special task 'to complement and where appropriate support the work of the EU–Russia energy dialogue at regional level with well-defined activities, such as studies, seminars and workshops and common activities'.[45] Examples of BASREC's efforts to strengthen the EU–Russia dialogue include consultations of national regulators and system operators to develop mutually acceptable tariff principles; a model of the synchronization of electricity flows; the efforts to facilitate access to networks; and the efforts of the BASREC working group on electricity to strengthen co-operation of transmission system operators and to compare the ways they organize their capacity reserves. BASREC's natural gas working group has also studied the effects of gas liberalization directives in the Baltic Sea region. Baltic Gas has formulated its priorities more generally by just insisting on the creation of a fair, harmonized, and predictable business environment.[46] Baltrel has proposed a three-stage programme linking the current state of the markets with the future possibility of establishing a pan-European energy market. First the prospects of reaching a common understanding will be analysed and common actions agreed. Then market development and their implications will be considered. Finally, a common understanding of topical issues directly linked to power market development is to be reached. At each stage new infrastructure and production capacities are to be developed. The emissions trading system is to be perfected and roles between all market participants better defined.[47]

The market strategies of energy companies demonstrate their preparedness for the formation of a pan-European energy market. Their business activities have two main directions. Russia's oil and gas companies in the upstream sector of production are interested in the transport and distribution sector of the consuming countries in order to test their ability to work there. The example of Gazprom going downstream into the three Baltic countries, Germany and Finland can be cited here. Lukoil has tried to acquire oil processing plants in Lithuania, Poland and Germany although with virtually no results so far. At the same time, German and Norwegian companies are interested in moving to Russia's upstream business.

The potential of a pan-European energy market is also evident in electricity trade in northern Europe. The gains ensuing from legal harmonization and infrastructure link-ups figured prominently in specific initiatives like the 2005–6 offer of Rosenergoatom to construct an additional electricity cable to Finland in order to enlarge electricity export to Finland and further to Sweden, and the efforts of Fortum to acquire a majority stake in the wholesale generation company being formed in northwestern Russia. However, both projects were blocked at the national level (see Chapter 6).

45 See 'BASREC GSEO Work Plan 2003–2005'.

46 For the objective of Baltic Gas, see <http://www.balticgas.org/Objective.htm>.

47 See 'Baltrel Working Agenda for the Period 2003–2005', <http://www.baltrel.org>.

On the whole, the steps taken by regional actors help to build an enhanced legal framework and facilitate the eventual formation of a pan-European energy market. BASREC concentrates on more short-term goals of legal approximation and enhancement of the legal framework, whereas business associations, in particular Baltrel, are more adventurous and long-term in their approach. It is also notable that companies seriously strive to act prior to the formation of a pan-European market by making cross-investments and by establishing joint ventures.

(iii) *ND energy co-operation as a platform for developing innovative ideas for the energy dialogue: implementation of the Kyoto Protocol.* The starting point here is that northern European countries differ widely in the amount of their CO_2 emissions (Figure 4.2).

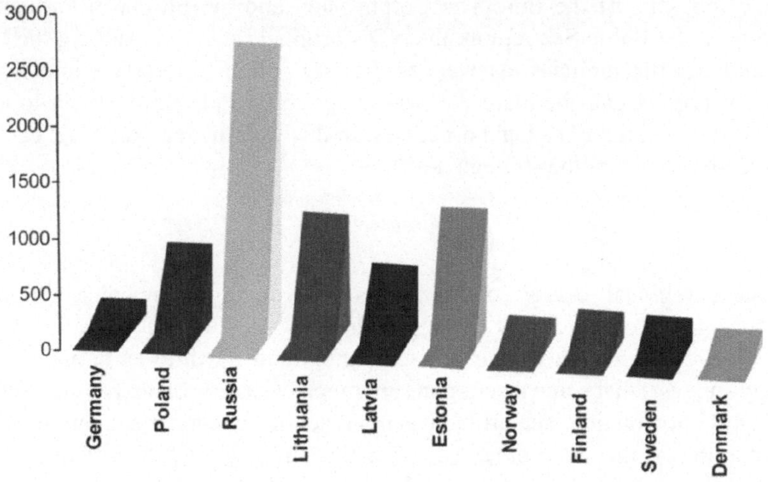

Figure 4.2 CO_2 emissions among northern European states (million tonnes)

Source: European Commission, 'A Northern Dimension for the Policies of the Union', communication from the Commission, Brussels, 25 November 1998, COM (1998) 589 final.

In 1999, in a conference in Helsinki, the ministers of energy of the Baltic Sea region suggested making the region a testing ground for international co-operation in the use of so-called flexible mechanism of the Kyoto Protocol. In the Vilnius conference in 2002, this concept was formalized and agreed to be carried out through BASREC. This paved the way for the energy ministers to agree with the European Commission on making the region a testing ground for joint implementation projects to reduce greenhouse gas emissions. Targets set for 2005 included the ratification of the Testing Ground Agreement by all the participants, guaranteeing the synchronization of policies with the EU, and more global activities in the field of climate change by involving business representatives. On the whole, the implementation of the Kyoto protocol, in particular its international flexible implementation component, is more advanced in the ND area than at the strategic level of the EU–Russia energy dialogue. In this sense, the example set at the regional level can well direct the further developments at the strategic level and point at potential problems.

ND co-operation can also alleviate the problems ensuing from the centrepiece of the Commission's concept for a pan-European energy market – Russia's proposed unilateral legal approximation with EU energy norms and rules without having a say on their present format or future development. The ND, by contrast, was initially conceived to involve Russia in EU decision-shaping. The ND can potentially engage Russia's representatives through 1) involving them in the preparation/consultation stage of new legislative acts at national and European levels, or 2) through the participation of Russian private actors in business associations, and through consultation between Russian and EU private sector actors, and in further position papers and lobbying of the European Commission and the European Parliament. Networking in the Baltic Sea region, the ND's original intent on bridging the EU's internal and external policies, as well as the ND's 2006 revamping into a joint EU–Russian project can facilitate the flow of policies and allow Russia to exert some influence over the EU's legal processes. In this way the regional level can also influence co-operation at the strategic level.

* * * *

In summary, regional energy organizations and energy companies supported actively the agenda of the energy dialogue. The discussed three cooperative modes of regional level interaction were characterized by their down-to-earth nature. This conversely suggests that some discussions are much easier to conduct at the strategic level than put into practice. The problem is not reaching an agreement, but the actual implementation. At the same time, it is clear that the potential of initiatives at the regional level deserves more attention for the simple reason that some of these initiatives can stimulate the overall development of the energy dialogue.

Several explanations can be found for the fact that most regional organizations in northern Europe support the EU's energy co-operation agenda. First, the majority of the countries in this region are EU members. For them, this makes co-operation along EU rules the only acceptable and available format. The same preference applies to EU based companies. The European Commission's membership in BASREC ensures conformity with *acquis communautaire* and EU policies. Second, energy companies support the EU's agenda because it promotes their activities and co-operates with them, eliminating investment risks and creating necessary conditions for their large-scale operation. From the Commission's perspective this is a way of providing stability to the consumers. Third, the EU's agenda was supported in the region rather than Russia's, due to Russia lacking any coherent long-term strategies other than the construction of export infrastructure in order to avoid dependence on transit countries.

Conflicting Patterns

The cooperative mode is not the only form of interaction between the strategic and regional levels. Although energy relations in the northern European region are down-to-earth and based on the interaction of a multitude of actors and on the legacy of many years of co-operation, certain issues go against the energy dialogue's agenda.

(i) *Preventing energy co-operation for reasons of national security of supply: development of transport infrastructure.* The most controversial project in regional energy relations is the German–Russian Nord Stream underwater gas pipeline. At the strategic level, the project is fully supported as a means of diversifying transportation of Russian resources to the EU area, and receives backing from Germany and the EU. Co-operation between Gazprom, E.On Ruhrgas and BASF went on regardless of the Baltic states' and Poland's vehement opposition (see Chapters 5 and 7).

One reason for the opposition is that the pipeline directly competes with the earlier designed two-trunk Yamal–Europe pipeline. Its completed trunk line links Russian energy resources with EU territory via Belarus and Poland whereas the construction of the second (parallel) line is postponed. The Amber line, which would have gone through Latvia, Lithuania and Poland, was subsequently abandoned as well. All these three latter countries are set to lose potential transit revenue when the Nord Stream pipeline is put into use. The historical and political experience of the Baltic countries and Poland plays a huge role in their priorities in the sphere of infrastructure. Being highly dependent on gas and oil supplies from Russia, they want to maintain their energy security by keeping Russia dependent on transportation through their territory. This reasoning is typical of Polish experts.[48] Polish, Lithuanian and Latvian authorities went as far as to argue that in the Nord Stream question, the German Government violated solidarity between EU members. Environmental uncertainties and possible risks were also cited as reasons for opposing the Nord Stream pipeline (see also Chapter 6). Similarly, Russia wanted to decrease the transit countries' influence. Some senior Federal Government officials still perceive them as exercising negative policies towards Russia.

Somewhat different conflicts characterize the efforts of Russian companies to acquire oil refineries in former members of the Soviet bloc. Polish and Lithuanian opposition to Lukoil acquiring stakes in their enterprises is particularly pronounced, and is fuelled by fears of increased Russian influence. Such sentiments can be viewed as barriers to any, even gradual transfer to a pan-European energy market. At the same time, Transneft's decision to temporarily cut oil supply to Lithuania's Mazeikiu oil refinery in 1999, allegedly because of the state of oil pipelines, yet coinciding with the sale of the refinery, did not facilitate mutual understanding. Instead of promoting the interests of Russia's companies in increasing its stakes in the processing and downstream market, it nourished fears of Russia's imperialism and prompted accusations of blackmailing by Russia's state company Transneft in order to facilitate the bid of Russian oil companies.

48 Agata Loskot, 'Security of Russian Gas Supplies to the EU – the Question of Infrastructural Connection', Policy Brief, Centre of Eastern Studies, Warsaw, February 2005.

In the Nord Stream and the Polish and Lithuanian cases, national governments are responsible for the conflicting mode of interrelationships between the strategic and regional levels; and in both cases they cited reasons of energy security. But arguably their concerns are also based on their historical experiences and the perception of a zero-sum game rather than the multifaceted realities of present-day co-operation.

(ii) *Preventing energy co-operation for reasons of protecting national markets: questions of legal framework and business environment.* An obvious example where regional patterns contradict the agenda of the energy dialogue is the case of oil transportation through the Baltic states' ports, particularly Latvia's Ventspils. First Moscow increased discriminatively railway tariffs for transporting Russian oil through Baltic ports, and as a result the Russian oil companies diverted oil traffic to the newly built Russian port of Primorsk (see also Chapters 7 and 8). Then transportation through the pipeline leading to Ventspils was stopped. This instance represented Moscow's effort to arbitrarily change the established rules, resulting in a serious distortion of competition.

Another example is the Estonian power plant in Narva, which refused to buy oil shale from northwest Russia on the grounds that it would result in Estonia spending its Kyoto protocol emission quotas for producing electricity from Russian oil shale. The Estonian Government and the electricity company Eesti Energia required Russia to either transfer its emission quotas to Tallinn or pay compensation. The real reason for the dispute was the EU directive on emissions trade from 2005, and the consequent Estonian realization of the greater profitability of selling emissions into the EU market than spending them on reprocessing Russia's oil shale.[49] This is, however, in a direct contradiction with the logic of the energy dialogue. Moreover, when emission quotas were originally allocated, Estonia and Russia were parts of the Soviet Union. The later redivision of the emission quotas took into account the fact that Russian oil shale was processed in Estonia.[50] Interestingly, some managers of Estonian enterprises admit that Russia would have won if the case was appealed in court. But the official policy of the Estonian Government was only a short-term compromise, eventually leading to the economic collapse of a whole mining town in Russia's northwest and then reorientation of its production.

Yet another example is the decision on the exploration in the Shtokman gas field that Russia announced in October 2006. After long and exhaustive negotiations

---

49 Elizaveta Dobkina, 'Zhertva Kioto', *Expert: Severo-Zapad*, 27 June 2005; 'Koncern Eesti Energia dolzhen poluchat pribyl za obrabotku rossiiskogo slantsa – sovetnik ministra ekonomiki Estonii', *Baltic News Service*, 12 July 2005; 'Parnikovyi effect', *Kommersant*, 7 May 2005; 'Peregovory rossiisko-estonskoi rabochei gruppy po slantsevomu krizisu zaversheny', Regnum, 11 July 2005; 'Estonskaya storona ozvuchila pretenzii "Leningradslanzu"', Regnum, 25 May 2005.

50 'Valerii Serdukov prigrozil Estonii lisheniem kvot', *Nezavisimaya gazeta*, 20 June 2005; 'Parlamentarii Lenoblasti prinyali obraschenie k presidentu RF i k rukovoditelu Evropeiskoi Komissii po povodu peresmotra kvot mirovym soobschestvom ot Estonii v polzu Rossii', Vse novosti Sankt-Peterburga, 25 May 2005; 'Rossiya mozhet podnyat vopros o lishenii Estonii chasti kvot po Kiotskomu protokolu – gubernator Leningradskoi oblasti', *Baltic News Service*, 10 June 2005.

with US, French and Norwegian bidders on the joint exploration of the gigantic gas field up in Russia's north, Gazprom announced that none of the proposals was good enough for a full partnership (see also Chapters 3 and 6). According to Gazprom's key officials, the company had been looking for concomitantly obtaining stakes in other exploration projects as well as in the down-stream business, and not so much for technologies or financial resources. This claim was fully in line with Russia's strategy announced during its 2006 G8 Presidency, that access to Russia's resources should be balanced by granting access for Russian companies to the distribution sector and by guaranteeing security of demand. At the same time, the decision undermines the good co-operation established in the region, especially in relation to this project. A parallel example where Russia's partner in the EU side prevented Russian actors' entry is the case where Rosenergoatom's project of constructing an extra electricity cable from Russia to Finland was blocked in 2006 (see Chapter 6).

These are examples of actions undertaken by state-bound companies in an alliance with their national governments in order to protect the economic interests of specific enterprises, at the expense of realizing a key element of the energy dialogue, i.e. enhancing the legal framework and business environment of energy co-operation.

(iii) *Preventing energy co-operation as a result of disputes with transit countries: security of supply.* Long-term security of gas supply has recently become crucial. This is due to the fact that Russian gas has to pass through transit countries such as Belarus and Ukraine on its way to the EU. The reliability of these transit countries is not certain because of their continued counting on cheap supply of Russian natural gas while refusing to sign agreements with higher prices. This situation led to a conflict with Belarus in February 2004. Gazprom halted the supplies of natural gas for transit for 18 hours, provoking Minsk to unauthorizedly take gas. As a result, industrial consumers in Poland, Lithuania and Germany suffered. The story repeated itself with some variations in early 2006 with Ukraine. When Gazprom stopped supplying natural gas to Ukraine in the absence of an agreement on prices, the Ukrainian company Naftogas started using transit gas destined for major EU consumers.[51] Many EU member-states, including Poland and Germany, missed a large share of their gas supplies. Thus Gazprom's conflicts with transit countries had serious implications for its EU partners particularly in northern Europe. Uneasy relations with transit countries were brought to light again in early 2007 when Gazprom pushed for the increase of prices in Belarus and for the transfer of the stakes in Belarusian gas pipeline system. This time the export of gas was not affected, but

---

51 The illegal siphoning of Russian gas by Ukraine was documented by SGS, an independent auditing company. Ukrainian officials refused to sign the measuring documents of the SGS.; see 'Kiev: Bez balansa gaza dlya Ukrainy ne mozhet byt rechi o tranzite', Rosbalt, 3 January 2006; 'Ukraina zayavlyaet, chto rossiiskii gaz ne voruet i ne potreblyaet', Korrespondent.Net, 2 January 2006; '"Naftogaz Ukrainy": My pobedili', Rosbalt, 4 January 2006; 'Germania zayavila, chto Ukraina voruet u nee gaz', Rosbalt, 3 January 2006; see also Nikolay Sokov, 'Alternative Interpretation of the Russian–Ukrainian Gas Crisis', *PONARS Policy Memo* 2006, no. 404. The Ukrainian crisis provoked a more visible reaction than the Belarusian one because it impacted several EU members. The political aspect of the crisis was also accentuated by its association with Ukraine's 'orange revolution' in 2005.

Minsk tried to take its revenge, at least a short-term one, by putting a new custom duty on the transit of Russian oil.

* * * *

To sum up, the conflicting mode of interaction between the strategic and regional levels is caused by either national governments or companies alone, or by their alliance. The reasons are divergent, but in most cases related to understandings of national or corporate interests, and to perceptions of the situation in zero-sum terms. In the case of Russia, they include efforts to maintain independence in energy exports and to ensure that energy security is provided to both consumers and producers by the symmetrical opening of the exploration and distribution sectors. In the case of the three Baltic countries and Poland, historical memory and fears of renewed Russian imperialism fuelled by energy stakes dominate. However, financial revenues and commercial positions are also considerably involved.

Conclusion

The analysis shows the substantial interaction between the northern European energy co-operation in the ND area and the strategic level of the energy dialogue. At the strategic level, the agenda of the energy dialogue has gradually come to include five broad issues: security of supply and trade development; transport infrastructure; enhancement of legal framework/investment climate and transfer of technologies; energy efficiency and Kyoto protocol, and establishment of a pan-European market. Due to internal EU reasons and Russia's insistence on solving short-term questions, the agenda of the dialogue is mainly technical. Even the more conceptual debates on a pan-European energy market are characterized by this technical nature. The European Commission dominates agenda-setting and the shaping of long-term relations as a result of its very specific long-term goals, whereas Russia until recently has tried to solve short-term problems, failing to come up with an independent vision on the energy dialogue's long-term development. In this way, the EU has ended up being the more strategic actor. However, Moscow's new assertiveness and promotion of energy co-operation based on equality of producers and consumers and reciprocal opening of exploration and distribution can gradually alter the situation.

At the regional level, the Russian side is mostly represented by the same actors that handle the strategic level dialogue. But in the EU side, the Commission's influence fades at the regional level. It limits itself to observing the conformity of multilateral relations with Community regulation, thus guaranteeing the agenda of the energy dialogue to be applied in the region. The EU's Baltic member states have taken a more active position, resulting in the continuation of old concerns and fears put into the energy agenda. Companies are more active at the regional level because it involves more down-to-earth co-operation. Regional energy co-operation involves also exclusively regional actors and organizations like BASREC, Baltic Gas and Baltrel, each with its own agenda.

The analysis demonstrates that interaction between strategic and regional levels has both conflicting and cooperative patterns. The overall balance is indisputably in favour of the cooperative mode. The dialogue is reinforced by the activities of regional organizations and companies. The northern European component is also based on the long tradition of energy co-operation in the area and the ability of regional actors to keep cooperating even in times of strained EU–Russia relations. At the same time, the conflicting patterns, arising mostly at the national (bilateral) level, cannot be ignored, particularly because they emerge from the actions of national governments of EU member states and some state companies. In some cases, the impulses come from the policies of Russia or its major energy companies. Similar instances can lead in the future to negative developments in the energy dialogue at the strategic level. Actions to neutralize them should address their national roots as well as the differences between the EU and Russia in how they see the long-term development of the energy dialogue.

Chapter 5

Germany and the EU–Russia Energy Dialogue

Kirsten Westphal

Introduction

This chapter analyzes the role of German policy and its impact on EU–Russia energy dialogue and EU–Russia energy relations in general. Germany has depended on Soviet/Russian energy deliveries for over thirty years. From the very beginning, energy trade was seen as a strategic element in the détente and co-operation with the Soviet Union, and the same perception carried through to relations with Russia.

This chapter first describes the long-standing role of Russia as a reliable energy supplier for Germany. This is something that is always taken up in the German energy policy narratives in order to justify closer ties with Russia. Second, the chapter discusses in more detail the actual content of German–Russian energy relations. German energy policy has for a long time relied on multilateralization of energy policies as an element of European integration. However, in the new millennium, German energy policies have accommodated specific business interests, whilst the energy related issues of climate change and environmental protection have been treated in close coordination with the EU. The German policy of securing its own close energy alliance with Russia has, to a certain extent, bypassed the EU–Russia energy dialogue, although both aim at reducing dependency risks. Therefore it needs to be asked whether the German approach has contributed to an enhanced energy security overall. Energy security demands a three-fold diversification in sources of energy, countries of origin, and transit, in order to contribute to a steady and secure energy supply at reasonable prices.

With the EU lacking a full competence in energy policy, and especially in foreign energy relations, the Schröder Government that ruled in Germany during 1998–2005 used its strong personal ties with President Putin's administration in Russia in order to promote German–Russian energy relations. It will be argued in this chapter that this weakened the EU Commission's strivings for a common foreign energy policy and limited the EU's opportunities to further diversify its energy supply and increase its energy security. Schröder's support for individual energy projects, such as the Nord Stream gas pipeline through the bed of the Baltic Sea, can also be explained by his government's strong ties with energy industry. Schröder's concomitantly close ties with Putin correspondingly benefited some German companies. But they also undermined a common European approach in external energy relations, and can be interpreted as legitimizing Putin's renationalization policies in the oil and gas industry. When Angela Merkel took office as the new German Chancellor in November 2005, she attempted to redefine the German–Russian relationship, at a moment when the Russo–Ukrainian gas dispute at the turn of 2005–6 impacted seriously on EU energy policies and the EU–Russia relationship in general.

Because natural gas deliveries represent the core of German–Russian energy relations, and because of the high investment needs to gas transport infrastructure, as well as the strategic long-term impact of decisions on transport routes, this chapter largely concentrates on gas. The policy recommendations of the chapter are: Germany should enhance the Union's role in energy policies *and* foreign policies; it should join in the recently bolstered efforts of diversifying energy supplies to German and European markets by proactively supporting infrastructure projects; and it should further integrate Russia into a multilateral set of common rules and regulations. At the face of these challenges, it will be concluded that the Merkel Government's record was patchy during its first year in office.[1]

The Historical Factor

In Germany, Russia traditionally enjoys a very good reputation for high dependability and delivery reliability since the so-called *Erdgasröhrengeschäft*, the natural gas pipeline deal wrapped up during the *Ostpolitik* era in order to economically back the political strategy of change through rapprochement (*Wandel durch Annäherung*). The building of first pipelines in the 1970s, against US objections and with billions of dollars of aid from western Europe, and particularly Germany, coincided with a political interest in a détente of the east–west conflict where Germany played a key role.[2] Chancellor Adenauer had visited Moscow already in 1955. Later the social democratic Chancellors Willy Brandt and Helmut Schmidt promoted *Ostpolitik* with the belief that better ties between Bonn and Moscow would ease tension in east–west conflicts and improve relations with East Germany.

Germany alongside other European states perceived energy deliveries from the Soviet Union as an opportunity to diversify energy supplies both in type of energy and geographical origin. In the European energy mix, natural gas was emerging as an important substitute for oil, and would also reduce dependency on the Middle East and OPEC oil. The new dependency on the Soviet Union was discussed in terms of the possible use of the 'energy weapon', because of the oil price shocks of the 1970s. At the end of the 1970s and in the 1980s, the aim of diversifying energy imports through increased deliveries from the Soviet Union was valued higher than the risk of greater exposure to Soviet pressure. Furthermore, as the deal comprised exchange of energy for technology, investments and highly manufactured goods from Germany, it was seen as a means of balancing the low German growth rates

1 The EU Commission's efforts are laid down in the new green paper on energy security; see European Commission, 'A European Energy Strategy for Sustainable, Competitive and Secure Energy', COM(2006) 105 final, Brussels, 8 March 2006.

2 Reimund Seidelmann, 'Die Sowjetischen Energieimporte in die Bundesrepublik Deutschland, Abhängigkeit oder politische Zusammenarbeit?', *Beiträge zur Konfliktforschung*, 1 (1982): 27–66.

by opening new export possibilities. The deal was thus explicitly reciprocal, and eventually mutual benefits created a web of interdependencies.

Natural gas imports from the Soviet Union to Germany started in 1973. The negotiations had begun in 1969, just before the German elections that resulted in the coalition of the Social Democrats and Liberals. The first gas deal was signed in 1970 along the lines of the Austrian deal signed in 1968, exchanging trade in natural gas to pipelines. The German deal encompassed credits from German banks to the Soviet Union to buy pipes from Mannesmann AG. The foreign currency receipts from the natural gas deliveries to Germany were used to pay back the credits. The main elements of this deal were the increased share of (Soviet) natural gas in the German energy mix, the creation of mutual interdependence through a deal including gas, credits and pipelines, and multilateralization by further supply to France, Belgium, the Netherlands, Austria and Italy as a risk reduction measure. For West Germany, according to Seidelmann,[3] the mutual interdependence connoted a worst case scenario of a total halt in gas deliveries, but the impact would be short-term because of alternatives, spare capacities, and storage capacities that had been built up. For the Soviet Union, the short-term effects would be low in a worst-case scenario, but would be grave in the long-term because of the 'imported' economic growth.

In summary, the gas pipeline deal added economic co-operation as an important element to the political détente by increasing structural and institutional co-operation. The political will of the Soviet Union to become a reliable and stable supplier was exemplified by its concession to include West Berlin into the agreement and deliver natural gas to the enclave, something which Bonn had for long been striving for. The Soviet Union turned out to be a reliable supplier which never used energy trade as a weapon in political conflicts, not even during the hot phases of the Cold War and momentary deterioration of the relationship in the beginning of the 1980s.[4] At that time, Germany faced harsh criticism from the US. The US had argued for the import of American coal instead of Russian gas. This, however, was neither economically nor politically attractive because of the interests of the German coal industry.[5]

Russia as the Main Supplier to the German Market

In the beginning of the 1980s, Soviet energy imports comprised only 7 per cent of the West German consumption.[6] This share is now significantly more at 30 per cent.[7] Oil imports from the Soviet Union comprised 2.9 per cent of the total imports in

3 Seidelmann, 'Die Sowjetischen Energieimporte', pp. 50–4.

4 In fact, German and Soviet leaders alike protested in 1981, when the US President Ronald Reagan imposed sanctions against the first big gas pipeline project from Russia to the West because of the declaration of martial law in Poland.

5 Seidelmann, 'Die Sowjetischen Energieimporte', p. 50.

6 Seidelmann, 'Die Sowjetischen Energieimporte', p. 53.

7 Author's estimations basing on statistics of Cerstin Gammelin and Fritz Vorholz, 'Sauber, aber explosiv', *Die Zeit*, 3 (2006), p. 20.

the 1980s,[8] and natural gas imports 17 per cent.[9] The import of Russian oil and gas has risen significantly since German reunification. The import of gas from Russia between 1991 and 2003 rose by 55.5 per cent, and oil imports between 1990 and 2003 by 57.4 per cent.[10] In 2005, Germany imported 38.3 million tonnes of oil from Russia. This translates into a share of 34.1 per cent of the overall imports. If the imports from Kazakhstan that go through Russian pipelines are included, this counts for 37 per cent.[11] For gas, the share of Russian imports is even higher at 41.7 per cent.[12] In terms of German consumption, Russian oil deliveries amount to 32 per cent of German consumption, and gas deliveries to 35 per cent due to Germany's relatively insignificant gas and oil production.[13] This means that the share of Russian hydrocarbons is overall very high, exceeding the 30 per cent limit recommended by the G7 as a critical import dependency indicator in 1983.[14] Even though short-term effects can be mitigated because of Germany's storage capacities amounting to 20 per cent of its annual consumption, mid-term and long term effects of disruptions in gas deliveries, or price surges, have to be seen as a potential risk.

Since the 1980s, Germany's dependence on Russia has not only risen in quantities, but has also changed in quality. The change in quality derives mainly from Russia's geostrategic aim of diversifying its export structure and agreements. Considering the high demand from China and India along with the growing concentration and limitedness of reserves, this can lead into a competition among consumers, giving Russia an extremely favourable position in setting the prices. This would relativize the European position as a major buyer of Russia energy. Moreover, the principle of reciprocity of treaties in the 1970s and 1980s – gas in exchange for technology (for the pipelines) – has been abandoned. This is especially important in the present context where Germany has to a large extent liberalized its electricity and gas markets and the same process is underway in the whole EU. Gazprom has exploited the opportunity to acquire shares, set up joint ventures, and to supply end-consumers, thereby expanding its 'mid- and downstream' activities into the German and the EU markets, and improving its value-added chain. Russia, however, has maintained a protected energy market favouring its own companies, and has used its gas pipeline monopoly as a bottleneck for alternative gas producers in the Caspian Sea basin and Central Asia who want to export their gas to Europe. In this way, Russia has significantly limited Germany's and the EU's possibilities of diversifying

8 Seidelmann, 'Die Sowjetischen Energieimporte', p. 40.

9 Seidelmann, 'Die Sowjetischen Energieimporte', p. 46.

10 Author's calculations based on Bundesministerium für Wirtschaft und Arbeit, 'Energiedaten, Nationale und internationale Entwicklung', Berlin, 31 January 2007, Table 13.

11 Roland Götz, 'Schweigen für Gas? Deutschlands Abhängigkeit von Rußlands Energielieferungen', *SWP Aktuell*, 43 (September 2004): 1–2.

12 Bundesministerium für Wirtschaft und Arbeit, 'Energiedaten, Nationale und internationale Entwicklung', Berlin, 31 January 2007, Table 17.

13 Author's calculations based on statistics by Gammelin and Vorholz, 'Sauber, aber explosiv', p. 20.

14 Friedemann Müller, 'Klimapolitik und Energieversorgungssicherheit: Zwei Seiten derselben Medaille', *SWP-Studie*, S 14 (April 2004): 23.

their energy supplies, and has secured its export and transport monopoly for energy flows from the east to the west. Western investments in the Russian energy sector have faced highly arbitrary circumstances (see also Chapter 3).[15]

The overall lack of equal and reciprocal access to markets and infrastructure in present EU–Russia energy relations is best evinced in Russia's non-ratification of the ECT treaty (see Chapter 1). And the relationship between individual EU members like Germany and Russia is also an asymmetrical one because private German companies have to deal with Russian state authorities and companies in which the Russian state has a majority stake, like Gazprom, compared to the government-to-government relationship of the past. The highly personal engagement of the Schröder Government that will be taken up in this chapter must in this light be understood as an effort to balance the existing asymmetries.

Alongside the growing structural asymmetries and the eroding principle of reciprocity, one must also bear in mind that Germany is Russia's most important trade partner, its biggest international creditor, and a major source for accumulated investments.[16] Germany ranks first in Russia's imports with 13.5 per cent originating in Germany, and second in Russia's exports with 8.2 per cent directed to Germany,[17] while Russia ranks only the 10th place for German imports, and 13th place for exports.[18] In 2005, imports from Russia were valued at 22.3 billion euros and exports at 17.3 billion euros, representing yet another increase from the previous years.[19] The underlying reason is the price surges for energy goods, which constitute the majority of Russia's exports (over 70 per cent). German exports, by contrast, are products of mechanical engineering, automobiles and parts, chemical products and electrical equipment.[20] But of course, oil and gas are not normal commodities, but highly strategic goods due to the dependency of all national economies and their growth rates on regular energy supplies and steady prices. And also to a certain extent, Germany's high dependency on Russian energy deliveries can be explained by Russia's significantly expanded role in the global energy market.

EU Energy Policies and the Russia–Germany Energy Relationship

EU energy policies became a determinant factor for Germany with the development of the single market in the early 1990s. On the one hand, there was a need to define a new governance structure for the internal European market. On the other hand, the new possibilities in the east prompted the EU to develop the ECT, the most

15 See Julia Kusznir and Heiko Pleines, 'Reichtum für wen? Politische Ökonomie rußländischer Ölfelder', *Osteuropa*, 9–10 (2004): 340–54.

16 Christian Meier, 'Deutsch-Russische Wirtschaftsbeziehungen unter Putin, Praxis-Probleme-Perspektiven', *SWP Studie*, 41 (November 2004): 5.

17 Goskomstat Rossii, 'Vneshnyaya torgovlya Rossiyskoy Federacii so stranami vne SNG', <http://www.gks.ru/free_doc/2006/b06_13/24–07.htm> (accessed 9 February 2007).

18 Statistisches Bundesamt, *Rangfolge der Handelspartner im Außenhandel der Bundesrepublik Deutschland 2005* (Wiesbaden: Statistisches Bundesamt, 2006), p. 1.

19 Statistisches Bundesamt, *Rangfolge der Handelspartner*.

20 Deutsche Industriebank, 'Russland Länderanalyse', p. 2.

encompassing multilateral regime in the energy field. This treaty reflects the euphoria of the end of the bipolar world in its pursuit of the idea of a pan-European energy market and to an extent picks up on the Soviet leader Gorbachev's idea of a common European House. Germany became actively involved in the ECT process, which was also valued as an effective instrument for German companies to increase their activities in Russia.[21] A German official, Peter Schütterle, became the secretary of the ECT secretariat. Supporting joint European efforts is the traditional multilateral approach of Germany towards European energy policy – after all, Germany and France originally formed the nucleus of the ECSC in 1951.

When Russia refused to ratify the ECT, mainly because of the ECT's Transit Protocol, the EU Commission initiated the EU–Russia energy dialogue (see Chapter 1). By this time, there was a growing awareness that the share of natural gas would increase in the future European energy mix because of its comparably low greenhouse gas emissions. Consequently, the then President of the EU Commission, Romano Prodi announced the so-called Prodi Plan in September 2000, according to which gas imports from Russia would be doubled to 240 billion cubic meters by 2020. Prodi's statement was planned to pave the way for a more concerted European energy policy.

This more concerted approach presupposes a consensus on policy goals, and more importantly, a transfer of decision-making power and sovereignty to the EU level. Neither a consensus, nor a streamlined decision-making process has been achieved. Sovereignty transfer in such a nationally sensitive policy field has also failed to materialize. Energy policy is still defined as a prerogative of the nation state. As a consequence, EU energy policy can only be conducted indirectly through other common policies in the fields of transport, environment and the single market (see Chapter 1). The differences in the member states' energy mix make this up to a point understandable, but the situation on the whole becomes problematic when large member states making use of different energy resources, and hosting powerful energy companies, either slow down or block decisions completely in the field of energy security. Although Germany used to rely on the multilateralization of energy policy, more recently it has approximated the energy policies of other large member states.

The EU Commission's attempt to start a debate on the security of energy supplies with the Green Paper of 2000 prompted adverse reactions in the member states that wanted to maintain energy policy as a national domain (see also Chapter 1).[22] Germany was no different. Since then, Germany has been hesitant towards the EU's energy dialogue with Russia, where the ECT and Russia's ratification of the Transit Protocol have recently gained renewed prominence, as they have in the G8. This all casts doubt on the consistence of Germany's approach. But in Germany the EU–Russia dialogue is widely viewed as the Commission's 'cup of tea'.[23] Taking Germany's position as a net importer of energy and a motor for EU integration,

---

21 Author interviews during 1996–8.

22 See European Commission, 'Green Paper – Towards a European Strategy for the Security of Energy Supply', COM (200) 769 final, November 2000.

23 Author interviews in 2005 in the Ministries of Foreign Affairs and Economic Affairs.

it is important to ask why Germany neglects or even side-steps the Commission's initiatives. This helps us to assess the prospects for the Commission's new efforts as manifested in its 2006 Green Paper that called for an enhanced dialogue with energy producers, first and foremost Russia, on the basis of a 'true partnership'. Yet, perhaps anticipating the reactions of Germany and other large member states, the 2006 Green Paper and the energy package of January 2007[24] do not mention the EU–Russia energy dialogue explicitly, but instead call for a new energy partnership with Russia that should be 'integrated into the framework of EU–Russian relations due to replace the current EU–Russia Partnership and Co-operation agreement in 2007'[25] and that should 'be based on market principles and those of the Energy Charter Treaty and draft Transit Protocol'.[26]

The Political Economy of the German Energy Sector and the Impact of Liberalization

The liberalization of the EU energy market has had a decisive impact on the political economy of the German energy sector. Liberalization started at the end of 1990s, and has since then been a source of dispute between member states and the Commission, as the scope and speed of implementing the Commission's directives depends on the member states. The liberalization of the energy market has to be seen as a by-product of the creation of a single market, and crucially, it was initiated during a time when energy prices were very low and not related to security concerns as now.

In Germany, the liberalization of the gas and electricity markets in 1997 and 1998 coincided with the change of government. The coalition of the Social Democrats (SPD) and Greens took over in October 1998. The following years witnessed German energy policy riding on two strategies simultaneously: on the demand side, the SPD–Greens coalition backed and implemented EU policies, including the Kyoto Protocol, related environmental policies, measures for increasing energy efficiency, facilitation in the use of renewables, and others. The SPD–Greens coalition thus joined the strong consensus on environmental and climate issues dominating in the EU. On the supply side, Germany emphasized the pursuit of national interests more openly than in the past. This thematic and strategic split between EU and German policies was reproduced at the institutional level when the competencies in energy policy were divided between the Ministry of the Environment and the Ministry for Economic Affairs. Germany's lack of energy ministry has been criticized explicitly.[27]

Although ties between German energy industry and the state have always been close, the late 1990s changed the quality of the relationship. The Energy Act came into force in April 1998. It was prepared under the old Kohl Government. The

24 European Commission, Communication from the Commission to the European Council and the European Parliament, 'An Energy Policy for Europe', COM (2007) 0001 final, Brussels, 10 January 2007.

25 European Commission, 'A European Energy Strategy for Sustainable, Competitive and Secure Energy', p. 15.

26 European Commission, 'An Energy Policy for Europe', p. 23.

27 In fact, nine of 14 ministries somehow deal with energy issues in Germany.

liberalization of energy markets in Europe implied a restructuring of property rights, the abolishment of territorial monopolies, and new rules to assure priority access for so-called green energy onto the electricity grid. This led to a transformation in the German state's role with the de-regulation and re-regulation for ensuring non-discriminatory access to networks in order to bolster competition. The shares owned by the German state, the federal states, and the local authorities are largely privatized. That Germany is the biggest and most attractive market, resulted in an exposure to pressure by foreign companies. France, the other important European market, has been reluctant to restructure and liberalize, but French companies are engaged in the German market. Therefore, Germany soon started emphasizing the necessity of reciprocity and equal access to markets within the EU and between the member states.[28] Comparable frameworks and ordering principles were seen as important preconditions for providing all companies with the same starting position. This served to slow down Germany's own liberalization. The end result is an oligopolist market dominated by four electricity companies, among them E.On and RWE. For many reasons, the initial approach of the government in implementing the EU directives was to rely on informal agreements aided by the close links the Schröder Government established with the energy industry.

The Schröder Government's coalition partner the Greens demanded a nuclear power phase-out. This spawned the so-called chancellor's round tables arranged with large energy companies in January 1999. The round tables constituted not only a lobbying group, but also a dense network. The first Minister for Economic Affairs of the SPD–Greens coalition, Werner Müller, a non-party expert, was the manager of a former electricity monopolist Veba that later merged with E.On. Schröder's contact with Veba and Müller date back to his time as the Prime Minister of Lower Saxony. It is said that in one of these round tables a plan to build a huge, competitive German energy company was conceived.[29] The merger between E.On and Ruhrgas was started in 2001, and was first rejected by the Federal Cartel Office, but was then approved in the Ministry for Economic Affairs by Alfred Hacke, a loyal ally of Schröder and Müller, in 2003 through a so-called minister approval, underlining the advantages for the overall national economy.[30] Germany, which never before had had a large multinational energy company, backed the merger of E.On and Ruhrgas to create an important and powerful player in both regional and international electricity and gas markets. At the same time, large energy companies shifted their corporate strategies away from utility functions like water supply in order to concentrate on the energy business per se.

The oligopolistic structure of the electricity and gas markets is a constant source of political disputes, in particular at the face of price increases for consumers and the large profits earned by energy companies. The non-transparent price structure, lack

---

28 See Werner Müller, 'Energiepolitik für die Zukunft', in Jürgen F. Baur (ed.), *Die Energiewirtschaft im Umbruch, Rechtliche und politische Probleme im neuen Wettbewerb* (Baden-Baden: Nomos, 2000), pp. 17–23.

29 See Cerstin Gammelin, 'Die Erdgas Connection', in *Die Zeit*, 51 (2005), 15 December 2005, p. 4.

30 Gammelin, 'Die Erdgas Connection'.

of real competition, and pressure from Brussels resulted in the new Energy Act that became effective on 1 July 2005, assigning the task of regulating Germany's electricity and gas markets to the Federal Network Agency to ensure non-discriminatory third party access to networks and to police the use-of-system charges levied by market players.

The politically supported merger between E.On and Ruhrgas meant far-reaching inroads into the energy market in order to secure the competitiveness and involvement of a German company in international energy trade. A wave of bids emerged ahead of the full opening of EU's energy market in July 2007,[31] as energy companies tried to consolidate their bases by mergers and acquisitions, and by maintaining and/or opening new production sites and trade relations in the EU and beyond. An earlier consolidation wave took place at the end of the 1990s, and what is evident in both waves is a classical conflict of goals: the goal of enhancing competition in the German market is in conflict with the goal of improving the competitiveness of German companies at the EU and global level. Imposed competition at the national level may reduce the competitiveness of companies outside the national economy, especially if the principle of reciprocity is not uniformly applied, and if the same norms and rules do not apply to all companies. This is particularly important in the electricity and gas sectors where cost-extensive networks are needed for transportation and distribution. As implied above, re-regulative interventions are consequently required on the national level, as illustrated by the German Energy Act of 2005, and the Union level, as proposed by the EU Commission.

Energy companies thus face competition on the national and EU levels, and in the more global context they compete not only with other multinational companies, but also with state-owned companies benefiting from a protected position in their domestic markets. In the German–Russian case, Gazprom is a state-controlled monopolist in the domestic market, with the Russian state having a share of 51 per cent, and simultaneously operates as a global actor in production, distribution, marketing, transport and export, and uses its thus acquired strategic weight to access the liberalized German and European markets.[32]

The policy of the Schröder Government during 1998–2005 can without a doubt be characterized by a shift away from a multilateral approach, towards a more unilateral pursuit of interests. This is well seen in the prevailing narrative that managing external energy dependency first of all demands powerful and competitive companies at the international level. Second, energy security becomes defined as demanding state responsibility for instance in the realization of large infrastructure and exploration projects abroad and can thus not only be left in the hands of private companies. This German policy has also impacted EU energy policy.

---

31 One of the most visible examples is the bid of E.On for the Spanish company Endesa.

32 See Kirsten Westphal, *Russische Energiepolitik: Ent- oder Neuverflechtung von Staat und Wirtschaft* (Baden-Baden: Nomos, 2000), pp. 61–3; 98–103; Andreas Heinrich, *Globale Einflussfaktoren auf das Unternehmensverhalten* (Münster.: LitVerlag, 2004), pp. 133–140.

German Energy Companies and Co-operation with Russia

Co-operation between the German Ruhrgas AG and the Soviet Soyuzgazeksport started in 1969. The German Mannesmann AG was also heavily involved in the deal because it held a monopoly in the production of special seamless giant pipes. The present-day E.On Ruhrgas, the major German gas company which imports around 60 per cent of Germany's overall gas consumption,[33] hence has a longstanding tie with Russia's gas industry, and strengthened it by increasing its stake in Gazprom in several stock acquisitions since 1998.[34] Currently, E.On Ruhrgas is Gazprom's largest foreign shareholder with direct and indirect interests in accumulation of 6.4 per cent of shares. The manager of Ruhrgas, Burckhardt Bergmann is a member of Gazprom's Board of Directors since 2000. Ruhrgas has contracts through 2030 to import 20 billion cubic metres of gas from Gazprom annually, which presently makes up about 30 per cent of the total imports of Ruhrgas.[35] The partnership between E.On Ruhrgas and Gazprom moves towards a bilateral monopoly between a producer and a buyer/re-seller, dominating the German market and flexing its muscles beyond. Reliance on Russian gas supplies via pipeline has so far withheld plans of building a liquefied natural gas (LNG) terminal in Germany. This leaves Germany the only one among the six largest industrial countries without an LNG port.

E.On Ruhrgas has a strategic interest in maintaining close energy relations with Gazprom, and in expanding into the exploration and production of Russian gas. This is a pressing issue for E.On Ruhrgas, because it produces only five per cent of the gas it sells, and strives to increase this share from 15 to 20 per cent.[36] In turn, Gazprom has a strategic interest in the further expansion into the German transport, distribution and retailing market. As a result of these ties, strategic deals are undertaken as quid-pro-quo-deals realized in exclusionary, non-transparent rounds of bargaining. For instance, the two companies signed Memorandums of Understanding in July 2004[37] and April 2005[38] that envisaged the joint development of the west Siberian gas field Yuzhno Russkoye (with a 24.5 per cent share for Ruhrgas), in return for Gazprom's participation in E.On's projects of gas and electricity transport and distribution in the German and EU mid- and downstream sector. To push through this package deal, Gazprom constantly presses E.On Ruhrgas for interests in its distribution subsidiaries, most notably Ruhrgas. So far, E.On has refused and instead has agreed to swap assets in three Hungarian companies for a stake in the Yuzhno Russkoye gas

---

33 'Gazprom setzt Eon unter Druck', in *Süddeutsche Zeitung*, 27 March 2006, p. 21.

34 E.On Ruhrgas AG, Chronik 1999 <http://www.eon-ruhrgas.com/deutsch/unternehmen/chronikg/1999.htm> (accessed 31 January 2006).

35 Judy Dempsey, 'For Many Countries, Nowhere Else to Turn', *International Herald Tribune*, 5 January 2006, p. 3.

36 'Gazprom setzt Eon unter Druck', *Süddeutsche Zeitung*, 27 March 2006, p. 21.

37 E.On Ruhrgas AG, 'Der Online Jahresbericht 2004', <http:onlinegb.ruhrgas.de/Highlights/Highlights_00.htm> (accessed 31 January 2006).

38 E.On Ruhrgas AG, 'E.On und Gazprom konkretisieren Schritt zu einer vertieften Zusammenarbeit', <http://www.eon.com/de/presse/news-show.do?id=637> (accessed 31 January 2006).

field, prompting harsh criticism in Hungary.[39] The Memorandum of Understanding of 2005 also mentioned the Nord Stream gas pipeline project through the bed of the Baltic Sea.

For German companies, the natural market for expansion is Eastern Europe. German electricity companies such as E.On and RWE are highly involved in Poland, Czech Republic, Slovakia, Hungary and Romania. E.On Ruhrgas and its subsidiaries, and RWE's subsidiary Transgas, have expanded into Poland, the Czech Republic, Slovakia, Hungary, Romania and the Baltic States, among others. In geographical terms this represents an effort to reach out to the Russian market through transporting Russian gas to Western Europe. The structural alliance with Gazprom, as exemplified by the asset swaps between E.On and Gazprom, BASF Wintershall and Gazprom, and their joint venture Wintershall Erdgas Handelshaus,[40] has not only implications for the new member states and their limited opportunities to diversify their energy supplies, but results in something close to a bilateral monopoly. This puts German energy companies in a key position in the EU market.

E.On Ruhrgas is not Gazprom's only German partner. Gazprom has co-operated with Wintershall (fully owned by BASF) since 1990. Wintershall has long-term gas delivery contracts with Russia until 2030.[41] Wintershall became directly involved in gas production in west Siberia through the founding of its joint venture Achimgaz with Gazprom to exploit a field near Novy Urengoy in 2003. In a package deal with Gazprom, BASF and E.On Ruhrgas build the Nord Stream gas pipeline through the Baltic Sea, with Wintershall obtaining a stake in the Yuzhno Russkoye gas field in West Siberia.[42] This is another significant step for Wintershall into the Russian production sector. The field has a volume of 500–700 billion cubic meters, equalling the German gas consumption for more than five years.[43] The Yuzhno Russkoye deal illustrates the institutional asymmetry within the German–Russian relationship. German companies have to deal with and rely on a monopolistic company in the Russian market, while Gazprom is able to bargain with both E.On Ruhrgas and BASF/ Wintershall. The strategy of Gazprom on the Russian market is to accept E.On Ruhrgas as the major foreign stakeholder, and Wintershall as a major partner in direct exploration and production. In the quid-pro-quo deal with BASF Wintershall, Gazprom obtained further stakes in Wingas[44] and exploration fields in Libya.

---

39 'Gazprom setzt Eon unter Druck'; Andrew E. Kramer, 'Gazprom and E.On to Swap Assets', *International Herald Tribune*, 14 July 2006, p. 10.

40 Wintershall Erdgas Handelshaus, in which Gazprom own 50 per cent, sells gas in eastern Europe.

41 Dempsey, 'For Many Countries'.

42 Wintershall, 'BASF und Gazprom schließen wegweisende Vereinbarung zur europäischen Energieversorgung', <http://www.wintershall.biz/214.html?&L=0> (accessed 31 January 2006).

43 Judy Dempsey, 'Baltic Pipeline to Bring Russian Gas to Europe', *International Herald Tribune*, 12 April 2005, p. 4; 'Gazprom setzt Eon unter Druck'.

44 One of the major and most strategic BASF Wintershall and Gazprom joint ventures is WINGAS in which Gazprom owns 35 per cent and which operates a web of long distance gas pipelines in Germany.

To realize its opportunities and ambitions in the liberalizing EU energy market, Gazprom has launched a number of joint ventures with BASF Wintershall, and acquired shares in German energy companies such as VNG (5.26 per cent) and Erdgas Mobil GmbH (4.5 per cent).[45] Gazprom does not rely on a single partner. Gazprom's German activities are coordinated by the German daughter ZGG GmbH that was renamed in October 2006 as part of the media campaign into Gazprom Germania, of which Gazprom holds 100 per cent.[46] Through its holdings in Wingas and Wintershall Handelshaus Gazprom maintains close ties with BASF.[47]

In a series of actions during 2006 and 2007, Gazprom attempted to become more active in the German midstream and downstream sector e.g. by buying shares of RWE, entering the retailing market and by engaging in the building of large storage facilities. The surprising coup to sponsor the German first league soccer team Schalke 04 from January 2007 on, in order to improve the damaged image after the 2005–6 Russo-Ukrainian gas dispute, can be seen as a clear first step into this direction. Experts estimate that Gazprom could double its income by deepening its value-added-chain and delivering gas directly to German consumers.[48]

Fragmentation between the liberalized EU market and the protected Russian market creates an asymmetric situation in which German companies face obstacles and experience setbacks, despite the high requirements for investments to keep the Russian energy sector working over the next 25 years, estimated by the IEA at approximately 330 billion euros (cf. Chapter 2). E.On Ruhrgas sees Russia's immense need for foreign funds and technology as an opportunity to 'play a role in the game' according to the CEO of E.On.[49] A notable case at this plane was the planned takeover by Siemens of Silovye Mashiny, a producer of equipment for power stations. The takeover was prevented by the Russian Federal Anti-Monopoly Commission. This was seen as a means of protecting vital Russian interests from foreign hands.[50] The Russian electricity monopoly, the Unified Energy System (RAO UES), later took over Silovye Mashiny. With the protectionist pressures in Russia, the personal relationship between Schröder and Putin worked to assure German companies a privileged position and helped to sign the Nord Stream gas pipeline and other major projects in German–Russian summits.

45 VNG , the former East German gas monopoly supplies the around 90 per cent of the east German gas market and has contracts with Gazprom lasting until 2012. Erdgas Mobil gmbH is a joint venture of gas suppliers that aims to expand the network of gas filling stations in Germany.

46 <http://www.gazprom-germania.de/german/company/konzern.php> (accessed 10 February 2007).

47 See <www.faz.net/s/Rub560251485DC24AF181BBEF83E12CA16E/Doc~ED7815B> (accessed 2 December 2005).

48 Hans-Willy Bein and Wieland Kramer, 'Bundesregierung fürchtet Gazprom-Pläne', *Süddeutsche Zeitung*, 7 November 2006, p. 19.

49 Judy Dempsey, 'Russian Energy Sector Needs Foreign Funds', *International Herald Tribune*, 19 May 2005, p. 5.

50 Daniel Brössler and Markus Balser, 'Rückschlag für Siemens in Russland', *Süddeutsche Zeitung*, 14 April 2005, p. 19.

To a certain extent, the structural linkages between German and Russian companies created an interest alliance that undermined, or at least weakened, the EU Commission's efforts to extend competition beyond EU borders. For instance, Wingas joined Gazprom in a meeting with the then EU Energy Commissioner de Palacio in April 2004 to discuss access to the European market. Both companies probably argued for maintaining the system of long-term contracts with duration of 15–20 years.[51]

The Schröder Factor: Towards a Russian–German Alliance?

The story so far is that the Schröder Government established closer ties with energy companies during the liberalization process and the negotiations for Germany's nuclear phase-out, and that it backed the merger of E.On and Ruhrgas in order to create a strong, powerful and competitive company which was thought to be in Germany's interest. The long-standing ties between E.On Ruhrgas and Gazprom enabled expanding their activities at the international level. There are good reasons for this pro-active German stance, but it also entails major shortcomings at the EU level.

Chancellor Schröder cultivated a special German–Russian relationship, and his personal friendship with President Putin during his term in office was seen as a guarantee for close energy relations. The personal ties between Schröder and Putin balanced off the selective and arbitrary legal circumstances hindering foreign investments in Russia. The outcome represents strategic energy projects based on German–Russian interests at the expense of the EU–Russia partnership.

This assessment of the energy dimension in the German–Russian relationship applies also to the overall bilateral partnership where economics has taken over politics. The primacy of economics has influenced EU–Russia relations due to Germany's decisive role at this plane. Under Schröder, Germany became the driving force for a renewed, special EU–Russia partnership. Not by chance, in the EU summit in Cologne in 1999, the EU approved the first-ever common strategy towards Russia. The Schröder Government functioned, to a certain extent, as an interlocutor in the ensuing new rapprochement between the EU and Russia, moving economic co-operation and trade matters higher on the agenda. One sign was the EU's acknowledgement of Russia's market economy status in 2002. Russia pushed for this for a long time in order to start negotiations on joining the WTO. The EU–Russia agreement on the Union's support for Russia's WTO membership bid in May 2004 was a package deal in which Russia ratified the Kyoto protocol and extended the PCA to the new member states, but rejected further reforms in the energy sector, other than gradually giving up the double-pricing for domestic and foreign buyers (see Chapters 1–4). Whether this really was part of a deal between Schröder, Chirac

51 'Gazprom, Wingas and EU Discuss Access to Energy Market', *Alexander's Gas and Oil Connections*, 9/8 (2004).

and Putin, or not, is an open question,[52] but at this plane it is interesting that Schröder has not publicly demanded any restructuring of Russia's monopolies.

At this point, the story justifying the German–Russian partnership under Schröder and Putin can be related to economic benefits and personal loyalty. The energy supply from Russia to Germany made an important contribution to German energy security. Schröder's policy also reflected a primeval social-democratic concern: the well-being of the German economy, its export capacity and strong companies. Within the German–Russian networking, interest-driven personal contacts yielded in 2000 the establishment of a special and high-ranking 'working-group for strategic issues of the German–Russian co-operation in economic and financial affairs'. This working group is unique in its composition and reflects the privileged nature of the partnership.[53] One observer sarcastically commented: 'Schröder the chief lobbyist of the stock corporation "Germany" is striving for a second function as an important trade missionary of Russia.'[54] This comment hints at how domestic (economic) interests determined German policy under the Schröder Government and emphasized its relationship with Putin's Russia. This all translated into a 'Germany first' strategy. In the EU context, the Schröder Government formulated German interests more bluntly than before, for example in the case of Germany's net contributions to the EU budget; and in the United Nations, Germany asked for a permanent Security Council seat.

The effects of the primacy of economics over politics, and the interest based policy turned a blind eye to the political shifts in Russia. Despite criticizing his predecessor Kohl for his 'sauna-friendship' with President Yeltsin, Schröder soon developed a close relationship, even friendship with the German speaking President of Russia. Schröder was consequently accused of maintaining silence in exchange of Russian gas.[55] He emphasized his personal influence on his close friend and refrained from criticizing in public the Russian policy in Chechnya, structural reforms in Russia, or the Yukos affair and the Khodorkovsky case (see Chapter 3). The then foreign policy spokesman argued that Schröder always raises these issues 'but between four eyes only' and that '(h)e knew what he could say privately and what publicly.'[56] Schröder also explicitly backed Putin in the 'fight against international terrorism' in Chechnya, and in many occasions supported Putin's argument 'that the international community has to come to a more differentiated assessment on Chechnya';[57] and stated that Putin was simply 'reconstructing statehood'.[58] This all prompted harsh

52 Pekka Sutela, in a meeting 'How Much Russian Energy is Good for Europe', 10 November 2005, <http://www. Carnegieendowment.org/events/index.cfm?fa=print&id=829> (accessed 25 January 2006).

53 Meier, 'Deutsch-Russische Wirtschaftsbeziehungen unter Putin', p. 18; for steps towards further institutionalization, see pp. 17–20.

54 Ulrich Schäfer, 'Schröders Russland AG', *Süddeutsche Zeitung*, 20 January 2005.

55 Roland Götz, 'Schweigen für Gas?'.

56 Judy Dempsey, 'Schröder and Putin Cementing Relations', *International Herald Tribune*, 8 September 2005, p. 3.

57 *Die Zeit*, 27 September 2001.

58 Daniel Brössler, 'Gas-Gigant in den Händen des Kremls', *Süddeutsche Zeitung*, 12 December 2005, p. 2.

reactions from the German opposition, the media and even Schröder's coalition partner the Greens. The decisive importance of personal contacts was affirmed by Schröder in an interview, where he said that he 'never had an incentive to distrust Putin.'[59]

The German–Russian partnership, defined as being of strategic interest to Germany, was reassured through a personal relationship. Moreover, both Schröder and Putin seemed to share enthusiasm for a close gas alliance which corresponds with the interests of Gazprom, E.On Ruhrgas and Wintershall. The new millennium brought about a broad consensus in Germany on relying on Russian gas and oil deliveries.

The Nord Stream Gas Pipeline

The Nord Stream gas pipeline – or North European Gas Pipeline, as it was named until October 2006 – is the crown of the personal relationship between Schröder and Putin, and as such, it works to cement German–Russian energy relations. The Nord Stream deal was signed deliberately on the eve of the early elections in Germany on 18 September 2005, and was interpreted as Putin's support for Schröder's campaign.[60] The Nord Stream pipeline will allow Gazprom to deliver gas from Vyborg directly to Greifswald on the north-eastern coast of Germany (see Figure 5.1). With a length of 917 kilometres, its first trunk line is set to be completed by 2010 with a capacity of 27 billion cubic meters. The gas can also be transported to France and the UK. The pipeline's second phase foresees spur lines to Sweden and the Russian exclave of Kaliningrad. The amount of gas will then increase to 55 billion cubic meters annually. The cost of this ambitious project will be between 6 and 10 billion dollars. Gazprom holds the majority stake of 51 per cent. BASF Wintershall and E.On Ruhrgas initially got 24.5 per cent each. Yet, allegedly Gazprom continued negotiations for including the Finnish Neste, French Gaz de France and the Dutch Gasunie.

The pipeline deal through the Baltic Sea is problematic for three reasons, and can only contribute partially to German and EU energy security. First, Russia's share in German gas imports can at times rise up to 40 per cent.[61] This can be a cause of concern. The technical shape that the Nord Stream project is taking relativizes these fears somewhat, but largely fails to bolster the project's potential otherwise.

59 Interview with Gerhard Schröder, 'Ich habe nicht die Absicht, die Russland-Politik zu verändern', *Süddeutsche Zeitung*, 2–3 October 2004, p. 7.

60 Daniel Brössler, 'Gas-Gigant in den Händen des Kremls', *Süddeutsche Zeitung*, 12 December 2005, p. 2.

61 *TAZ*, no. 7764, 9 September 2005, p. 3.

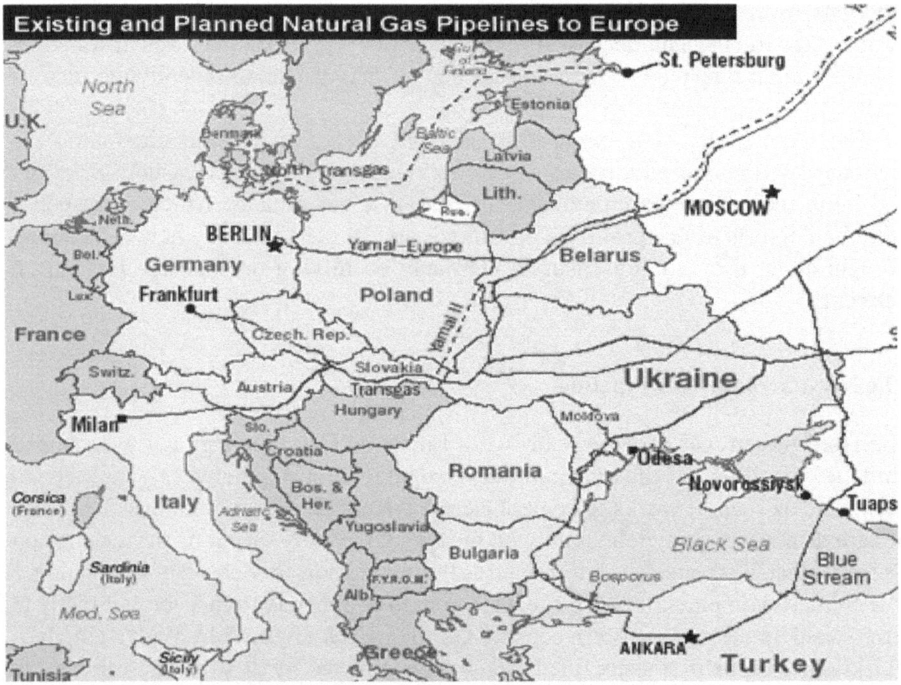

Figure 5.1 The Nord Stream gas pipeline

Source: Energy Information Administration; the Nord Stream gas pipeline is marked by the striped line in the Baltic Sea.

The Nord Stream pipeline was originally planned to run from the giant Shtokman field in the Barents Sea by a pipeline system to be built through Finland and Sweden and then along the Baltic Sea bed to Denmark and Germany. The EU gave the project the status of a Trans-European Network Project. Despite increased British interest in Russian gas, the EU never spent any money on a feasibility study.[62] When Gazprom decided in 2004 to use LNG technology in the Barents Sea, this pipeline project became less attractive for northern European countries, and the Nord Stream pipeline became mainly a bilateral Russian–German issue.

However, in October 2006 Gazprom and Putin announced that the Shtokman field, which is among the world's largest untapped gas fields, would be exploited by Gazprom solely without relying in the planned American, French and Norwegian consortium partners. With this cancellation Gazprom returned to the original idea of using gas from Shtokman to fill the Nord Stream pipeline (see also Chapter 6). With the Nord Stream running, German gas imports are expected to increase from 87 billion cubic meters to 100 billion cubic meters by 2020 according to BASF, and

62 For further detail, see Roland Götz, 'Die Ostseegaspipeline – Instrument der Versorgungssicherheit oder politisches Druckmittel?', *RusslandAnalysen*, 72 (2005), pp. 12–15.

as high as 150 billion cubic meters by 2030. The imports of the EU–25 are expected to double to 450 billion cubic meters a year by 2015 according to Ruhrgas.[63] In the beginning, the pipeline will be filled with gas from the west Siberian Yuzhno Russkoye field. It is doubtful that Russian deliveries from this field will meet the increasing demand.[64] Gazprom's reversion to using gas from Shtokman must be seen in this context. However, the Russian energy strategy also foresees diversifying exports to the east and south, and Gazprom has planned a number of large, expensive and complicated projects such as the Nord Stream and the Altai Pipelines, and the exploration of the Yamal gas fields that bind huge sums of investment. There is a big question mark vis-à-vis the parallel realization of these projects, in particular vis-à-vis a parallel putting-on-stream of Yamal and Shtokman. EU buyers will have to compete with other buyers, especially for energy sources located in these remote areas. This will raise gas prices and call for a concerted, multilateral effort from EU member states. At this plane, the second phase of building links from the Nord Stream pipeline to Sweden and Kaliningrad becomes of strategic importance.

The imperative for multilateralization hints at the second problematic dimension of the Nord Stream pipeline. For Gazprom, this project is a priority because it will help significantly in reducing Gazprom's problematic dependence on its traditional transit routes through Ukraine, Belarus and Poland (see Chapters 2, 3, 4 and 8). This makes the pipeline an important bargaining chip enabling Gazprom to put more pressure on the transit countries with regard to transit fees and gas prices, leaving the transit countries with little leverage because of their dependence on Russian energy. The other fear raised within the transit countries is that the amount of gas transported through an alternative trunk system will be reduced in the future, in favour of a larger amount being exported through the Nord Stream pipeline. Therefore, the Nord Stream project is interpreted as a threat to the transit states' energy security, transit fees and energy prices' stability. Poland, an EU member country, as well as Ukraine, have raised strong objections. Poland and the Baltic states also criticized their omission from the planned links to the Nord Stream and lack of consultation by Russia and Germany.[65] Moreover, the pipeline through the Baltic Sea carries ecological risks of leaks, which, deep in the sea, take time to be discovered and identified. The Nord Stream pipeline is on the whole seen as an economically costly, essentially political project, especially since the alternative pipeline routes to Germany and Western Europe are at the moment operating below full capacity.[66] Modernizing and reconstructing them would have been cheaper, and alternative projects also exist.

The most important trunk system from Russia, the *Bratstvo*, runs through Ukraine and Slovakia to western Europe, supplying between 70 to 80 per cent of Germany's Russian gas imports, and almost 90 per cent of Europe's gas supplies from Russia

63 Judy Dempsey, 'Schröder Gets a Pipeline Job (and Criticism)', *International Herald Tribune*, 10–11 December 2005, p. 8.

64 See in more detail Götz, 'Rußlands Energiestrategie und die Energieversorgung Europas', *SWP Studie*, 6 (2004).

65 Dempsey, 'Schröder gets a pipeline job'.

66 Jonathan Stern, quoted in Judy Dempsey, 'Gazprom's Dream Pipeline to Europe', *International Herald Tribune*, 14 March 2005, p. 1.

overall.[67] This pipeline has a capacity of 175 billion cubic meters a year and is currently transporting 115 billion cubic meters. This means that an increase of gas imports of 60 billion cubic meters a year could be achieved through the installation of new pumping and compressor stations on the transport route. This would cost less than the Nord Stream pipeline project. Chancellor Schröder, President Putin, and the then President of the Ukraine, Leonid Kuchma, agreed in June 2002 to build a consortium to reconstruct and modernize the Ukrainian export trunk line.[68] This consortium never agreed on the working conditions. Gazprom made several attempts to take over the control of this export pipeline system, the latest taking place in the run-up to the Russian–Ukrainian gas dispute in December 2005–January 2006. In part, these were attempts to cope with Ukraine's illegal siphoning of gas, but geostrategic considerations were involved as well (see also Chapter 4).

The Yamal–Europe pipeline offers another alternative to the Nord Stream pipeline. It was originally planned to link the rich gas reserves of the Yamal Peninsula with Western Europe through Belarus but bypassing Ukraine.[69] It was intended to have two trunks with a total capacity of 60 billion cubic meters a year. But its popularity faded, mainly because Gazprom shifted its priorities towards the Nord Stream pipeline. As a consequence, the second trunk has not been built and the pipeline has not been linked to the Yamal Peninsula, where the gas fields remain unexploited. Currently, the pipeline exports gas with a capacity of 28 billion cubic meters a year to the EU after a new compressor station was installed. In 2007, as a result of the Russian–Belarusian energy dispute on gas prices and transit (see Chapters 3 and 4; also below), Gazprom managed to take over shares of Beltransgaz, the company that manages the Belarusian part of the Yamal pipeline. The Nord Stream pipeline questions the realization of the second trunk of the Yamal pipeline, for which one of the plans was to construct it running south through Ukraine, and the other, the Amber project, through Latvia, Lithuania and the Kaliningrad *oblast'*, linking into the Yamal–Europe pipeline in Poland.[70]

The third problematic dimension derives from the fact that Schröder was taking the job of a supervisory board chairman of the Nord Stream pipeline company soon after leaving his post as the German Chancellor and that Matthias Warning, a former Stasi officer, became managing director. What made the scandal even more pronounced in Germany was the fact that it became public in April 2006 that Germany stands surety for a credit of two German banks to Gazprom for the construction of the Nord Stream pipeline.[71] This declaration of surety was approved in October only a few days before Schröder was handing over to Merkel. Leaving aside Schröder's

67 Loskot, 'Security of Russian Gas Supplies to the EU', p. 1.

68 See Götz, 'Die Ostseegaspipeline'.

69 Astrid Sahm and Kirsten Westphal, 'Power and the Yamal Pipeline', in Margarita M. Balmaceda, James I. Clem and Lisbeth Tarlow (eds), *Independent Belarus: Domestic Determinants, Regional Dynamics, and Implications for the West* (Cambridge, Mass: Harvard University Press, 2002), pp. 270–301.

70 Agata Loskot, 'Security of Russian Gas Supplies to the EU – the Question of Infrastructural Connections', Policy Briefs, Centre for Eastern Studies, Warsaw, p. 2.

71 Ulrich Schäfer, Daniel Brössler, Hans Leyendecker and Nikolaus Piper, 'Ein völlig normaler Vorgang', *Süddeutsche Zeitung*, 4 April 2006, p. 5.

damaged image, what makes his step highly problematic is the evident interpretation that his policies serve to legitimize Putin's energy policy record, if not explicitly then implicitly. In several steps, the Putin administration regained control over significant parts of Russia's oil and gas industry (see Chapters 3 and 8). These moves were criticised by its former economic advisor, Andrei Illarionov, as state-corporatism.[72] The re-nationalization took place at a time of growing competition among energy consuming countries looking at Russia as the main producer outside OPEC. Russia's energy sector is characterized by selectivity and inequality before law. State control has cut into political and economic freedoms. The approval of Gazprom's export monopoly by the Russian *Duma*, the Yukos affair and Sibneft's 'merger' with Gazprom exemplify this political strategy.

This assessment of the Nord Stream gas pipeline project leads to a highly problematic picture. But looking from a strictly German perspective, it is, first of all, a direct link from Russia to Germany without any transit countries. Second, it follows that it helps to diversify transport routes to Germany and reinforces the mutual interdependence between the buyer and seller. Third, Schröder being a supervisory chairman represents a guarantee of the project's continuance. Finally, the project without a doubt strengthens the position of German companies E.On Ruhrgas and Wintershall on the international and EU markets as gate keepers to the EU market.

The Impact of German Energy Policies on the EU–Russia Energy Dialogue and EU Energy Policy

Germany has a central position as the main entrance gate for Russian gas into west European markets. However, Russia's natural gas production will become more cost-extensive with the west Siberian gas fields depleting. Russia's own consumption is also likely to increase with all the consequences for export volumes. The very cold winter in 2006 presaged well the very potential supply bottlenecks. Add to this the number of supply contracts Russia has signed with other consumers like the US, China and Japan, and the need to diversify supplies becomes evident. Germany is one of the largest energy consumers within the EU, and should take a special interest in energy policies at the EU level and in the energy dialogue with Russia. The risks in the Russian direction call for Germany to diversify its supplies from other countries and regions, also including the decision to build an LNG port in Wilhelmshaven. Such an approach should be harmonized with the other member states and conducted at the EU level. The existing EU internal energy market also demands this.

The German shift towards bilateral energy policies has detracted from a multilateral approach, and in the process it has abandoned the Union level attempts of building institutions and markets.[73] The EU aims at managing *interdependence* with the help of universal and equally applied rules as evinced already in the ECT. The Union

72 Andrei Illarionov, 'When the State Means Business', *International Herald Tribune*, 26 January 2006 (summary translation from *Kommersant*).

73 See Clingendael International Energy Programme, *Study on Energy Supply Security and Geopolitics*, final report prepared for DG TREN, The Hague, January 2004.

also relies on market mechanisms in its aims at moderating access to resources, whilst simultaneously and strongly involving private companies. But the German emphasis on a bilateral German–Russian strategic alliance, Germany's preference for concrete energy projects, and the strong state involvement have been conducive to more traditional geopolitics in the wider regional market. The resulting effect on international governance structures cannot be neglected, because the more energy producing or consuming countries follow such a geopolitical path pursuing a neo-realist, state security centred approach, the more difficult it is to set up institutions for multilateral co-operation grounded on legal institutions and market mechanisms. This is significant due to the fact that international energy politics depends a lot on the roles played by states, as they are the ones keeping issues of sovereignty, political authority and territorially defined power on the agenda. For them energy resources are a power resource both on the national and international level.

Simultaneously the German approach and the personal relationship cultivated by Schröder with Putin failed to account for a *real* strategic partnership defined by common norms, rules, institutions and shared values. The relationship was limited to a close interest alliance characterized by very concrete projects. Deeper and broader German–Russian alliance relationship would require the formulation and embodiment of an all-encompassing political, economic and societal project going far beyond the existing web of economic, trade and financial relations. For this end, open and fair debate is necessary, but hidden criticism disadvantageous.[74]

Although Germany's energy relations with Russia are based on a traditional logic and have a longer history than the EU's energy relationship with Russia, the outcome in both cases has failed to follow the major principles of managing energy dependency and securing energy supplies: diversification and multilateralization. From this perspective, the focus should be opened to the EU market. The new EU member states depend on Russian imports for more than 73 per cent. Some are almost completely dependent on Russia (see Chapter 1).[75] For these countries, diversification can only mean getting more supplies from the EU market, which, for its part, will have to increase and diversify supplies from Norway, the Caspian Sea/Central Asia and North Africa, through alternative trunk lines such as the Nabucco project and through new LNG terminal facilities. There is an argument for bilateral energy ties, but even if this may have increased German energy security, the problem is the wider *geostrategic* impact of Germany's exclusive energy relations. Germany's exclusive alliance has to a certain extent undermined a common EU approach, or the multilateral approach of the ECT that would integrate other energy abundant countries of the Caspian Sea and Central Asia.

A more multilateral approach would take the interests and fears of the new EU member states into consideration, unlike was done for example in the Nord Stream pipeline project. Although the Nord Stream has a multilateral dimension of sorts with regard to old member states in Western Europe by strengthening Germany's position as a gate to the wider EU market and by having some spur line plans on

74 Hannes Adomeit, Katrin Bastian and Roland Götz, 'Deutsche Russlandpolitik unter Druck', in *SWP Aktuell*, 56 (2004): 8.

75 Loskot, 'Security of Russian Gas Supplies to the EU', p. 4.

board, it was perceived in the new member states as dividing EU energy markets. Polish newspapers went as far as to compare the Putin–Schröder pact to the Molotov–Ribbentrop pact agreed on the eve of the Second World War.[76] Even if this is exaggerated, the factual outcome for these countries is loss in strategic leverage as transit countries with an eye on negotiations on energy prices and greater exposure to gas delivery cut-offs. Operating with an argument on diversification from the exclusive Russian energy supplies for the new member states through Germany and the rest of the old member states could have cooled down the protests. But this never occurred to the Schröder Government.

The fact that Germany gave primacy to economics over politics may have contributed to a depolitization within the EU and helped to de-couple economic and political issues. Normally it is the Council of Europe or the Organization for Security and Co-operation in Europe (OSCE), that attack loudly problematic political issues such as the human rights situation in Chechnya, the Yukos case, or restriction of the media in Russia, whereas the EU Commission and the EU Council prefer to concentrate on day-to-day technical issues.[77] The 'orange revolution' in Ukraine and the rallying of new EU member states like Poland behind the Yushchenko alliance of revolutionaries in Ukraine also politicized the EU–Russia relationship. The German Government, by contrast, maintained silence, and left criticism to the then opposition party Conservatives. Earlier, the close Schröder–Putin–Chirac alliance against the US led war in Iraq in 2003 had already divided the EU and led to a rift between Germany and the US. Schröder built the anti-war alliance at the expense of transatlantic relations, and in the process helped to point at divisions between 'old' and 'new' Europe. All this resulted in the emergence of a cleavage between the old, big and powerful member states and the new member states. The Nord Stream project only added more fuel to the woes of the new members. In these conditions, multilateralism within EU energy policy was restricted to climate and environmental policies.

The New Merkel Government: Readjusting German and EU policies?

Upon Merkel's taking of office in November 2005, various signs appeared for a readjustment of German energy policies, also with regard to the EU. This can be mainly explained by two events that served as a real *ruptura* of the traditional German storyline of reliable energy supplies from Russia: Gazprom's dispute with Ukraine in 2005–6 and the Russian–Belarusian case in the beginning of 2007. During the 2005–6 dispute, Gazprom reduced the flow to Ukraine 20 per cent, emphasizing

76 Jan Pinski, 'Pakt Putin-Schröder: Gazowe okrazenie polski', *Wprost*, no. 1189, 18 September 2005 <http://www.wprost.pl/ar/?0=80802> (accessed 28 March 2006).

77 See Kirsten Westphal, 'The EU–Russian Relationship and the Energy Factor: A European View', in Kirsten Westphal (ed.), *A Focus on EU–Russian Relations: Towards a Close Partnership on Defined Roadmaps?* (Frankfurt am Main et al.: Peter Lang, 2005), pp. 1–36.

that it expected Ukraine to maintain the gas flow westward at unchanged levels.[78] However, EU member states reported reduced gas deliveries. During the Russian–Belarusian dispute, Russia cut off the Druzhba oil pipeline (see Chapters 2, 3 and 4).

Merkel also readjusted German policies towards Russia. She is not only the first woman in office, but also the first head of government from East Germany. Whilst fluent in Russian, she is said to have an ambivalent perception of Russia. As the leader of opposition, Merkel criticized Schröder for promoting the interests of German industry at the expense of human rights issues, particularly regarding Chechnya and attacked him for 'double standards' due to Schröder's criticism of the Bush administration and concomitant silence on Putin's policies. Merkel has emphasized freedom as an important principle and has publicly announced that she views developments in Russia as a 'cause of concern'.[79]

Merkel made her first official visit to Moscow shortly after returning from Washington in January 2006. She thus followed in the footsteps of the former conservative Chancellor Kohl, who always tried to balance Germany's relationships with the US and Russia. Yet, there are many signs of Merkel's policy having nothing in common with the policy style of her male predecessors. For example, she talks to representatives of human rights organizations. This on its own marks a shift in German policy. The Russia policy was laid down in the coalition accord, stating that Germany wants to develop a strategic partnership both at the bilateral level and EU level.[80] Whilst Merkel redefines the policy towards Russia in questions of human rights, rule of law, Chechnya, and with regard to the EU's neighbourhood policy, the Minister of Foreign Affairs, Frank-Walter Steinmeier, who is the former head of Schröder's Administration, represents continuity to the Russian side. That Merkel has played a very strong and cooperative role in the EU, is partly due to the weakness of other European heads of government, and partly a conservative tradition established by former Chancellors like Adenauer and Kohl.

The Russian–Ukrainian crisis took place two months after Merkel took office in November 2005. Prompted by the crisis, the new German Government voiced the first loud doubts of the dependency on Russian imports, notwithstanding the fact that the coalition accord puts an emphasis on a long-term energy partnership with Russia.[81] Fingers were also pointed at the high involvement of the Russian president. This is an important shake-up in German politics. The German Minister for Economic Affairs, Michael Glos, openly raised doubts about the responsibility and reliability of Russia and made a plea for diversifying Germany's energy supplies, for instance by going back for nuclear power plants. Energy policies moved to the top of the German political agenda, and consequently the Chancellor organized a

78 Dan Bilefsky and Carter Dougherty, 'As Supplies Drop, EU Officials Appeal to Moscow', *International Herald Tribune*, 3 January 2006, pp. 1, 15.

79 'Merkel Criticizes Jails like Guantánamo Bay', *International Herald Tribune*, 9 January 2006, p. 3.

80 CDU/CSU/SPD 'Koalitionsvertrag zwischen CDU, CSU und SPD: Gemeinsam für Deutschland – mit Mut und Menschlichkeit', 11 November 2005, p. 135.

81 CDU/CSU/SPD 'Koalitionsvertrag zwischen CDU, CSU und SPD'.

series of energy summits in Germany. Moreover, after Schröder's policy created the E.On–Ruhrgas merger, the Nord Stream gas pipeline project and other elements into the exclusive German–Russian network, the new German Government seems to proceed safely to strengthen co-operation with other EU member states. This has kick-started talks on energy security in Germany and the EU, but it will remain to be seen whether this results in profound reforms in EU energy policy.

The predominant narrative among European publics is that gas and oil deliveries to Ukraine and Belarus were used as a foreign policy lever by Russia. Leaving aside the other circumstances and implications of these energy crises (see Chapters 3 and 4), what is important here is the way it reinitiated a debate on energy security in the EU and particularly on the relationship with Russia, as the Commission's 2006 Green Paper on energy and the energy package of January 2007 illustrate. Under discussion is for example the issue of whether energy security can be left in the hands of private companies. In current practice, energy security is an issue for private companies, and in case of crises, member states assume a role. Only later on, if the crises cannot be solved the issue is passed to the 'gas coordination group' of the EU. A more assertive EU competence would imply substantial changes in EU energy policy. Only in 2005, member states watered down an energy directive and failed to assign Brussels control over storage, extra supplies and management of energy crises. So far, the Commission has been almost powerless when it comes to encouraging competition.

Germany took over the presidency of the EU council in the first half of 2007. Simultaneously Germany assumed the equally energy policy pertinent presidency of the G8 from Russia. In the run-up to Germany's chairmanships, there were various domestic debates on a readjusted relationship with Russia and on challenges to global energy security. With regard to the German–Russian and EU–Russian relationships, the government strongly promotes a new *Ostpolitik* with the idea of 'rapprochement through interdependence' (*Annäherung durch Verflechtung*) and aims to engage Russia in a web of mutual interdependencies with energy as a key component.

However, Merkel's policy for a more concerted EU approach in energy policy is ambivalent. Not surprisingly, Merkel criticized Spain and France for their protectionist measures at the face of the merger bids by E.On and Enel for local energy companies, emphasizing that the EU increasingly needs 'European champions' instead of 'national champions.'[82] Within the new debate on energy policy on the EU level, Germany favours, together with Britain, more openness in the energy sector and more diversification of energy sources. On the other hand, Merkel rejected, together with the Netherlands, the EU's attempts at exerting multilateral control over storage facilities in order to maintain exclusive use of Germany's cost-extensive but large gas storage facilities. In the same vein, she bypassed the issue of granting more

82 Dan Bilefsky, 'A New Round of EU Feuds', *International Herald Tribune*, 24 March 2006 <http://www.iht.com/articles/2006/03/23/business/energy.php> (accessed 20 March 2007).

powers to the EU and establishing a new regulatory authority on the EU level that would focus on the need of diversifying and opening energy grids.[83]

Soon after the Commission issued the 2006 Green Paper, Germany outlined its ideas on European energy policy, emphasizing co-operation with EU member states in energy efficiency and climate issues.[84] With regard to the external dimension, Germany emphasized the revitalization of the ECT, closer pan-European energy co-operation and 'a more powerful approach' by the EU and its member states in their external relations with energy producing and consuming countries.[85] However, it is the external dimension, particularly relations with Russia, where the EU is still split despite Merkel's announcement of bringing Poland and other eastern European states to the negotiation table for example to discuss the Nord Stream gas pipeline project. Poland is a pivotal case at this plane, because of the diverging positions of Merkel and Poland's PM Kaczynski on deeper EU integration, as well as their relations with Russia which have so far proven to be on very separate tracks. The cacophony in the EU over how to shape the EU–Russia relationship is so intense that the negotiations on a new PCA agreement have proved to be very difficult and have come to a halt at various points.

The latest developments illustrate what a long and winding road there is ahead for the Commission's energy package of January 2007. It is very likely that EU member states agree on the principles of sustainability, efficiency and innovation contained in the document, but their positions will diverge when it comes to unity, integration and solidarity.[86] This complicates the creation of a fully liberalized energy market, the implementation of solidarity, and last but not least, a more concerted external energy policy towards energy producers, transit countries and other energy consumers. What Katinka Barysch called the 'resurgence of economic nationalism with energy seen as a strategic sector', weakens the market position of the EU and its 450 million consumers significantly.[87] It is also doubtful that the Ukrainian-Russian gas and the Russian-Belarusian energy dispute will serve as '*radical* EU energy shake up',[88] or a catalyst for a more collective and effective EU energy policy. As Jonathan Stern puts it, it will require years to reach an agreement within the EU.[89]

---

83 Judy Dempsey and Dan Bilefsky, 'EU Unity on Power is Elusive', *International Herald Tribune*, 23 March 2006, <http://www.iht.com/articles/2006/03/22/business/energy.php> (accessed 22 March 2007).

84 Bundesministerium für Wirtschaft und Technologie/Bundesministerium für Umwelt, Naturschutz und Reaktorsicherheit, 'Energieversorgung für Deutschland. Statusbericht für den Energiegipfel am 3. April 2006', Berlin, März 2006.

85 Frank-Walter Steinmeier, 'Energy Security: Avoiding Conflict over Fuel', *International Herald Tribune*, 23 March 2006.

86 See Kirsten Westphal, 'Whither Europe? Energy Policy between Multilateral Governance and Geopolitics', *Internationale Politik und Gesellschaft*, 4 (2006): 44–63.

87 Katinka Barysch, cited in Judy Dempsey, 'EU Antitrust Chief Threatens Energy Suppliers with Investigations', *International Herald Tribune*, 17 February 2006, <http://www.iht.com/articles/2006/02/16/business/euenergy.php> (accessed 24 March 2007).

88 Dempsey, 'For Many Countries'; emphasis added.

89 Ibid.

Conclusion

Schröder's policy of giving primacy to economics over politics and his emphasis for a 'Germany first' strategy was designed in the midst of fundamental EU reforms in the Union's own energy market and the simultaneously ongoing globalization of energy policy characterized by three factors: 1) a remarkable power shift away from the consuming countries to the producing countries; 2) a relative loss in the (economic) power of EU member states because of the emergence of China, India and Brazil; 3) a realization of energy markets facing crises both in the demand *and* supply sides, which has revealed the necessity for huge investments over the next few years in (protectionist) energy producing countries.

In this environment, support for German energy companies to become competitive and expand their activities can be seen as a way of enhancing German energy security. However, this has negative implications for the EU's attempts of promoting multilateral governance and common institutions in the relations with Russia, and has given more weight to geopolitical approaches of structuring energy trade in general, and between EU and Russia in particular. In this light, the shared security of supplies fears within the EU that emerged forcefully during 2006–7 can be viewed as a small window of opportunity for reformulating EU energy policies and engaging Russia anew into a web of interdependencies.

The German Government is called to multilateralize its policies especially in relation to the new member states that are locked into long-term gas contracts with Gazprom.[90] These dependencies will make the EU critically dependent on Gazprom in the upcoming years. The imperative of diversification needs to be taken seriously. The share of Russian gas has already passed the critical threshold in Europe, especially because Russia's control of pipelines to Europe from the east and south-east through Turkey hinders and blocks diversification from the Caspian Sea. The Turkish market is already saturated with Russian gas pumped through the Blue Stream Pipeline running in the Black Sea.

To diversify supplies, Germany and the other member states have to deepen their co-operation in order to manage interdependence and achieve sustainability by multilateral means. The EU has to enhance its competence in regulating the EU-wide market and in external energy policy. Large infrastructure projects require political support, but have to take the needs of the EU market into account. With regard to the EU–Russia partnership, a new effort is needed to construct common spaces by using spill-over effects from energy relations. The talk should not be about a one-way take-over of EU legislation, or even a transfer of parts of the *acquis*. Such an approach has always been criticized by Russia for good reasons.[91] The aim should be at implementing fair and reciprocal access to markets and infrastructure, and the creation of equally applied, non-discriminatory and transparent rules for investments. Negotiations have to be conducted not only in the framework of the energy dialogue between the Commission and Russia, but also within the overall

90 Ibid.

91 Katrin Bastian und Roland Götz, 'Unter Freunden? Die deutsch-russische Interessenallianz', *Blätter für Deutsche und Internationale Politik*, 5 (2005): 583–92.

EU–Russia relationship with stronger support from EU institutions and member states. The definition of rules of the game is at stake and has to be realized in a reciprocal manner. This presupposes the political will and proposals from both sides on how to structure an *energy community* that can function as a basis for a sustainable strategic partnership building on co-operation and dialogue.

Chapter 6

The Nordic Countries: Engaging Russia, Trading in Energy or Taming Environmental Threats?

Pami Aalto and Nina Tynkkynen

Introduction[1]

This chapter elucidates the northern European regional level of the EU–Russia energy dialogue by looking at the Nordic countries. These states are unified by their strong legacy of historical ties and subsequent Nordic co-operation since the Cold War era. Simultaneously they have very divergent relationships with the EU, and in many respects, also with Russia. In energy trade questions, they display heterogeneity as well, ranging from Finland's high dependence from Russia to Norway's almost equal position as another major energy producer. The Nordic states' joint contribution to the EU–Russia energy dialogue is to strengthen the principle of sustainability. Here the Nordics derive from their tendency of promoting the sustainability principle both in bilateral and multilateral forums, even when they are fully aware of it not always being the preferred item on the agenda of other actors. In this way, the Nordic region elucidates further the country-specific challenges that the EU–Russia energy dialogue encounters at the regional level of the Northern Dimension area.

The starting point for an examination of the Nordic countries' role in the energy dialogue must be the mix of their mutual ties and differences. The group of Nordic countries or *Norden* – consisting of Denmark, Finland, Iceland, Norway and Sweden – are closely linked together by linguistic, cultural and historical ties.[2] Institutionally they linked by their co-operation in international organizations and within the inter-parliamentary organ Nordic Council and the Nordic Council of Ministers (NCM).

---

1 Pami Aalto bears the main responsibility for the first two sections after the present jointly prepared introductory text, and Nina Tynkkynen for the two final sections. However, both authors have commented on and edited each other's text.

2 In terms of culture, the Nordic states are linked together by the nineteenth century Scandinavianist movement and the close connections between the Danish, Icelandic, Norwegian and Swedish languages. Finland has a Swedish-speaking national minority. In terms of political history, the present Finnish territory was under the Swedish crown until the early nineteenth century and then an autonomous part of Tsarist Russia before achieving independence in 1918. The Norwegians were ruled by the Swedish and Danish crowns before Norway's independence in 1905. Iceland achieved independence from Denmark in 1944.

The deepening and widening of European integration has shaped inter-Nordic ties profoundly. Denmark joined the EEC in 1972, but its later opt-outs from the EU's Maastricht treaty in 1993 served to keep Denmark's EU integration market-based whilst remaining outside the single currency. Finland and Sweden joined in 1995, but Sweden also decided to stay outside the single currency. Norway remained outside the Union after a referendum in 1994. According to Janne Haaland Matlary, the Norwegians' suspicions towards EU energy policy which was emerging at the time was central to the 'no' vote, although Norway received special guarantees of maintaining sovereignty over its oil and gas resources.[3] Together with Iceland, today Norway is bound to the single market's energy sector related regulation and observes its development through membership in the European Economic Area (EEA), which, however, withholds any voting rights in decision-making. Yet, Norway has in many instances supported the EU's foreign policy positions.

Despite of their divergent ties with the EU, the Nordic countries are all bound to EU energy policy through Union or EEA membership. But they also co-operate among themselves in energy questions. They have established working parties within the NCM on electricity markets, climate policy and regional co-operation. In 2002, the Nordic Energy Research institution was founded to coordinate research in this field, with headquarters in Oslo. But interestingly enough, energy co-operation within the NCM largely parallels the co-operation within the umbrella of the Northern Dimension. Since its inception in 1997–8, the ND has been strongly EU-funded and EU-designed, and the availability of large amounts of funds from EU sources is pivotal for this.[4] The 2006 revamping of the ND to accommodate Russian, Norwegian and Icelandic concerns better may not, however, make the ND much more Nordic, as it was simultaneously re-modelled along the lines of the four EU–Russia 'common spaces' on the economy, external and internal security, and culture and education.[5]

The mix of the Nordic countries' mutual ties and differences in relation to the EU, and their energy co-operation and its links to the ND evokes questions of the extent to which the Nordics are in tune with the strategic level objectives of the EU–Russia energy dialogue. Taken that from the EU's part these objectives include facilitating energy trade and investment by means of providing a political and institutional framework, whilst at the same time trying to integrating member states' positions

---

3 Norway especially feared that the EU would develop common policies for concessions, rate of resource depletion, or for any other upstream aspects of gas production; see Janne Haaland Matlary, 'The Nordic Countries and EU Membership: the Energy Factor', in Lee Miles (ed.), *The European Union and the Nordic Countries* (London: Routledge, 1996). A competing explanation for Matlary's energy argument for Norway's 'no' to EU membership is the suspicions of EU membership endangering traditional livelihoods and lifestyles in Norway's fisheries and agricultural sectors; see Iver Neumann, 'This Little Piggy Stayed at Home: Why Norway Is Not a Member of the EU', in Lene Hansen and Ole Wæver (eds), *European Integration and National Identity* (London: Routledge, 2002).

4 Pami Aalto, *European Union and the Making of a Wider Northern Europe* (London: Routledge, 2006), chs. 2, 4 and 9.

5 European Council, 'Northern Dimension – Guidelines for the Northern Dimension Policy from 2007, 14358/1/05 REV 1, Brussels, 18 November 2005.

(see Chapter 1), this chapter asks: do the Nordics manage to introduce any specific building blocks for a regional level interface to the EU–Russia energy dialogue, or, is their insistence on combating environmental threats related to energy production and transportation in their own neighbourhood making dialogue more difficult?

This chapter first surveys in a comparative fashion the ability of the Nordics to engage Russia into a dialogue, their patterns of trading in energy, and their actions in taming the associated environmental threats. The policy narratives used by Norwegian and Finnish actors to justify their actions in the energy field are then selected into a more detailed analysis. The conclusions suggest that despite Finland, Sweden and Denmark are in various ways active in the development of the EU's relations with Russia, especially with regard to regional level co-operation, in the end, they remain somewhat ill shaped actors in the EU–Russia energy dialogue. Norway, by contrast, pays lip service to this dialogue whilst concomitantly holding its own energy dialogue with Russia on a different agenda and vocabulary. On the whole, it is suggested that in order to develop a pan-European energy policy more attuned to solving collective energy problems in the wider European area, the Nordics should be better involved into the EU–Russia energy dialogue, and that this inevitably means widening the dialogue's agenda to deal more explicitly with the environmental and other impacts of energy trade of concern to them.

Engaging Russia, Trading in Energy or Taming Threats?

The first general observation is that the Nordic countries have very divergent traditions and present-day patterns of engaging Russia. During the Cold War, Denmark belonged firmly into the North Atlantic Treaty Organisation (NATO) and thus represented the western camp opposing the Soviet Union, although it kept a low profile in the alliance. Denmark has a relatively thin historical tradition of relations with Russia, and is geographically distant from Russia's borders. However, in 1993 Denmark initiated together with Germany the Council of the Baltic Sea States (CBSS) to get the Russians and Balts into the same table with the other northerners, in order to deal with various 'soft security' issues ranging from humanitarian and health matters to environmental protection, energy and nuclear safety, and culture and education, to transport and communication. Denmark supported the ND by organizing a conference on Russia's Kaliningrad region during its chairmanship of the Union in the latter part of 2002. However, Denmark drove itself at odds with Russia by advocating loudly the Baltic states' NATO membership, criticizing Russia's handling of the conflict in Chechnya, and against Russian protests, gave asylum to Ahmed Zakayev, the representative to the now deceased Chechnyan leader Aslan Maskhadov, both of whom Russia calls terrorists.

Sweden has a distant historical legacy of great power rivalry with Russia. During the Cold War, Sweden stayed mostly neutral in its publicly presented policies, although it concomitantly found itself in diplomatic incidents with the Soviets, in particular during the 1980s when Soviet submarines were suspected of operating in

Swedish territorial waters.[6] Later on, the fairly close, covert military-political ties of Sweden with the west and NATO were made public. Sweden's EU membership further consolidated its active role in the various Baltic Sea regional organizations that emerged in the early 1990s. Sweden hosts the secretariat of the CBSS, which has a special energy unit providing assistance to the chairmanship of the Baltic Sea region's energy co-operation organization BASREC and its Group of Senior Officials (GSEO; see Chapter 4). Sweden also supports the ND activities.

The Norwegians in the northern part of the country have a distant tradition of interacting with Russians in the Murmansk region as part of the so-called Pomor trade – barter exchange of grain and fish in people-to-people contacts until the Russian revolution in 1918 sealed the Norwegian–Russian border.[7] Norway's NATO membership, despite the country's low profile in the alliance, hindered contacts during the Cold War. The collapse of the Soviet Union opened up a window of opportunity for Norway to initiate the multilateral forum Barents Euro-Arctic Council (BEAC). The BEAC also has its regional forum and conducts activities related to environmental and nuclear safety issues in the energy resources rich but heavily militarized areas east from Norwegian–Russian border, to the Kola peninsula and beyond in northwest Russia. Regardless of Norway's lack of EU membership, Barents co-operation represents its input into the ND, with the BEAC having become one of its implementation organs.[8] After the running out of the second ND action plan by the end of 2006 and with the revision of the ND, Norway became better incorporated.[9]

Finland tried to maintain its sometimes precarious independence during the Cold War by declaring neutrality within the constraints of the Finnish–Soviet Friendship, Co-operation and Mutual Assistance (FCMA) treaty. With strongly personalised leader-to-leader relations reminiscent of Germany's *Ostpolitik* (see Chapter 5) and gradually growing trust between the Finns and Soviets, the Soviet Union abstained from invoking the military consultation paragraphs in the FCMA treaty. Gradually Finland became a model case of co-existing with a bigger neighbour with a different system of production.[10] Since the 1970s, the Finnish–Soviet confidence building measures led to multilateral environmental co-operation via the establishment of the Helsinki Commission. At the bilateral level, Finnish–Soviet scientific co-operation complemented the HELCOM activities. In the 1985 framework agreement on environmental co-operation, the Finnish and Soviet governments recognized that transboundary pollution needs to be combated through mutual co-operation. In 1992

6 For a critical opinion on these suspicions, see Ola Tunander, *The Secret War Against Sweden: US and British Submarine Deception in the 1980s* (London: Frank Cass, 2004).

7 Geir Hønneland, 'Identity Formation in the Barents Euro-Arctic Region', *Cooperation and Conflict*, 33/3 (1998): 277–97.

8 Anne Myrjord, 'Governance beyond the Union: EU boundaries in the Barents Euro-Arctic Region', *European Foreign Affairs Review*, 8 (2003): 239–57.

9 European Council, 'Northern Dimension'.

10 Hans Mouritzen, 'Security Communities in the Baltic Sea Region: Real and Imagined', *Security Dialogue*, 32/3 (2001): 297–8; Tapani Vaahtoranta and Tuomas Forsberg, 'Post-Neutral or Pre-Allied? Finnish and Swedish Policies on the EU and NATO as Security Organisations', *UPI Working Papers*, 29/2000, pp. 8–9.

Finland and Russia signed an agreement on co-operation in the Murmansk Region, the Republic of Karelia, St. Petersburg and the Leningrad Region.[11] The agreement defined the legal framework for the so-called neighbouring areas co-operation of the Government of Finland and permitted financial support for areas adjacent to Finland.[12] In the process of achieving EU membership in 1995 and thereafter Finland put emphasis on multilateral work,[13] the flagship project being the ND proposed by Finland in 1997. An environmental partnership fund NDEP was launched for the ND in 2002, with capital set to climb to over 2 billion in total. Planning for a partnership in transport and logistics was started in 2005. The prospect of a separate energy partnership was also mentioned in connection to the ND's revamping in 2006.

Hence, during the 1990s, each Nordic country started bilateral programmes for engaging Russia and at the same time participated in multilateral regional co-operation frameworks launched by themselves and subsequently incorporated into the ND umbrella. In the new millennium, energy questions started to figure more prominently in these regional activities. However, the Nordics have very divergent energy relations with Russia. This leads us to the second observation on how divergent positions they occupy vis-à-vis the EU–Russia energy dialogue.

Denmark's oil and gas resources in the North Sea allow it to direct oil for exports and sell some gas through a pipeline to southern Sweden.[14] Recent expansion of gas and oil recovery in the North Sea fields has made Denmark energy self-sufficient, with exports exceeding domestic consumption by 139 per cent already in 1999. Although Denmark still remains a small player in energy exports, and lacks any major energy companies, its distinct character is the accent on sustainable energy production. From 1988 to 2000, energy use grew only by 2 per cent whereas the economy grew by 27 per cent, putting to rest the traditionally positive correlation between economic growth and energy use (see Chapter 2). Use of sustainable energy resources hit the level of 14 per cent in 2005, with the construction of a large wind mill park in the North Sea since 2002 helping to produce 27 per cent of electricity by renewable sources.[15] Denmark has initiated liberalization in its energy market to respond to EU requirements, and exercises high environmental taxes accounting for about one fifth of the Danish government's budget.[16] But despite Denmark

11 Agreement between the Governments of the Republic of Finland and the Russian Federation on cooperation in the Murmansk Region, the republic of Karelia, Leningrad Region and the City of St. Petersburg, 20 January 1992.

12 Although the agreement did not contain any obligations to reduce pollution, it facilitated opportunities for future environmental co-operation; see Heidi Hiltunen, *Finland and Environmental Problems in Russia and Estonia* (Helsinki: The Finnish Institute for International Affairs, 1994), p. 36.

13 For a more detailed analysis, see Nina Häyrynen, 'Changing Strategies of Environmental Co-operation between Finland and Russia', in *Northern Veche: Proceedings of the Second NRF Open Meeting* (Akureyri: Northern Research Forum, 2004), pp. 109–116.

14 Matlary, 'The Nordic Countries and EU Membership', p. 239.

15 European Commission, 'Overview: Energy for the Future: Renewable Sources of Energy', <http://ec.europa.eu/energy/res/index_en.htm> (accessed 12 March 2007).

16 See Royal Danish Ministry of Foreign Affairs, 'Environment and Energy: Catalogue of Danish Exporting Companies', 2002, <http://www.um.dk/NR/rdonlyres/F7EA3813-146E-

expecting a growing market within the wider European area to its sustainable energy technologies, it has remained a somewhat peripheral actor in the EU–Russia energy dialogue thanks to its self-sufficiency and lack of border with Russia.

Sweden's considerable nuclear power, hydropower and biomass generation has helped it to maintain a 30 per cent share of renewables and lower energy imports dependence than the EU average.[17] In 1991, corporate and political actors agreed to postpone the earlier, referendum-based decision on nuclear power phase-out until alternative sources are found. One means of substituting nuclear power is increasing gas imports. Although the share of gas in Sweden's energy mix was a negligible 2 per cent in 2004, it has rapidly taken a 20–5 per cent share of the market where it is available in the west coast through the pipeline from Denmark. There are plans for building LNG facilities, building a new pipeline from Germany via Denmark to Trelleborg, and eventually also for the creation of a Nordic gas market. Gas imports from Russia are projected in connection to the possible realization of the second phase of the Nord Stream pipeline from the St. Petersburg area via the Baltic Sea to Germany (see Chapter 5). Oil imports from Russia have increased notably during the new millennium, hitting 27 per cent of Sweden's oil imports in 2004, but the work of the governmental commission against oil dependence precludes further reliance on oil.[18]

A further means of linking Sweden more closely to Russia is the proposed project of building an electricity cable from Kernovo near the Sosnovyi Bor nuclear power plant southwest from the St. Petersburg area. One possible route for the cable goes through Kotka, Finland, and then through the existing Finnish–Swedish cable to Sweden, with Swedish end users buying most of the extra capacity. In summer 2006, the Finnish electricity grid operator Fingrid was at loggerheads over the project with Finnish industrial users who would snap up part of the offered extra capacity whilst expecting electricity prices to sink by 3–9 per cent.[19] Apart from the costs between 600 million and 1.4 billion euros estimated by the Finnish Energy Market Authority, underlying the heated debate were fears of the doubling of Russian electricity supplies in an already heavily Russia-dependent country in energy issues. Another factor was that the plan introduces the Russian state-owned nuclear power company Rosenergoatom into the Finnish markets through its subsidiaries United Power and Baltenergo. This would compromise the alleged plans of Fingrid's owners Fortum

4132-B991-BD563A35A6C4/0/eksportkatalog_uk.pdf> (accessed 29 June 2006).

17 European Commission, 'Overview'; Sweden's energy imports dependence rate was 45 per cent in 2005, whereas the EU–25 average was 56.2 per cent; see Eurostat, 'Energy in the EU: First Estimates 2005: EU25 Energy Consumption Equivalent to More than Three and a Half Tonnes of Oil per Capita', *Eurostat News Release*, 126/2006, 21 September 2006; Robert. L. Larsson, *Sweden and the NEGP: A Pilot Study of the North European Gas Pipeline and Sweden's Dependence on Russian Energy* (Stockholm: Defence Research Agency, June 2006), p. 39.

18 Larsson, *Sweden and the NEGP*, 39–43.

19 The estimate is calculated by the VTT, The Technical Research Centre of Finland; see Olavi Koistinen, 'Energiamarkkinavirasto tukee Fingridin laskelmia sähkökapelista', *Helsingin Sanomat*, 20 August 2006.

and Pohjolan Voima to import electricity from Russia independently.[20] Both economic and political arguments were thus in circulation. The offer to Swedish end users divided not only Finnish players but also Russian suppliers, with Rosenergoatom looking to supply its nuclear power generated electricity in relative independence from the Russian electricity exporter RAO Nord.

The Finnish Minister for Trade and Industry Mauri Pekkarinen eventually declined permission for the new cable in December 2006 by referring not only to the feared costs and technical uncertainties on the Russian side, but also to a need to become more self-sufficient in power generation.[21] The heated Swedish–Finnish–Russian cable episode shows the extent to which Swedish actors are slowly becoming more interested in Russian energy sources, and how other actors, in this case Finnish ones, almost immediately become drawn along due to the inherently interconnected and transborder nature of energy policy.

As a major energy exporter benefiting from its oil, natural gas and hydropower resources, Norway's energy production is more than 12 times its domestic consumption, 60 per cent of which is sustainably produced using mostly hydropower.[22] In 2003, Norway supplied approximately 14 per cent of the natural gas consumed in Europe, and 20 per cent of German imports and 30 per cent of the amount consumed in France. The new Ormen Lange field will deliver 20 per cent of the needs of the UK when taken into full use.[23] 90 per cent of Norway's oil and gas goes for export, and the petroleum sector accounts for 21 per cent of GDP.[24] In the positive long-term scenario suggested in the Norwegian Government's white paper in 2002, the country's oil resources would suffice for 50 years production and gas resources for 100 years. In the absence of investments into the remaining, not as easily exploitable

20 The lower figure calculated by the Finnish Energy Market Authority is higher than some of the earlier estimates by independent experts, and relates to strengthening the Finnish grid in order to deal with the projected increased supply. The higher figure incorporates the costs of shielding the Finnish grid with extra connections to Sweden in order to deter any threats on the grid's collapse on the Russian side. Although such cases are rare, in the harsh winter conditions of 2006 the Russian party had to cut its export supplies. United Power disputed any fears of grid collapse, and said it is willing to build extra capacity near Sosnovyi Bor, but failed to provide any firm timetable. United Power deemed the costs for building extra connections from Finland to Sweden an internal Finnish matter, and argued that there was only a 50 million euros investment need; see Koistinen, 'Energiamarkkinavirasto tukee Fingridin laskelmia'; Heikki Arola and Anu Lassila, 'Merikaapeli ei saanut lupaa kolmen vuoden jahkailun jälkeen', *Helsingin Sanomat*, 20 December 2006.

21 Heikki Arola and Anu Lassila, 'Merikaapeli ei saanut lupaa'.

22 Einar Steensnæs, 'The World Summit on Sustainable Development – The Petroleum Industry, Perspective and Response', speech by the Minister of Petroleum and Energy at the 17th World Petroleum Congress, 2 September 2002, Rio de Janeiro; 'Norwegian Energy Scene', opening speech by the Minister of Petroleum and Energy at the Global Energy Foresight Seminar, 26 May 2004, Stavanger.

23 Steensnæs, 'Norwegian Energy Scene'; Odd Roger Enoksen, 'Oil and Gas Offshore Developments in the Arctic and Cold Regions', speech by Minister of Petroleum and Energy, INTSOK seminar, 25 January 2006, Moscow.

24 Steensnæs, 'Norwegian Energy Scene'; Enoksen, 'Oil and Gas Offshore Developments in the Arctic and Cold Regions'.

fields in the Norwegian continental shelf that require the development of costly advanced technologies, oil production would virtually stop by 2020.[25]

Norway exports electricity to the other Nordic countries depending on the seasonally changing energy balance within the Nordic electricity grid NORD–EL that is moderated by the Nordic electricity market Nordpool. With the opening of the Estlink cable between Finland and Estonia in early 2007, further prospects of energy exports are opening up. Norwegian companies, mainly the partly privatized Statoil, and the partly state-owned Norsk Hydro, whose energy activities Statoil bought in December 2006, have traditionally operated mostly in the upstream part of energy flows.[26] But they remain interested in the further development of the European energy market and its regulation, especially as Norway has opened its own gas market and participates in the EU's internal energy market. In fact, the EU and Norway have their own energy dialogue with regular meetings since 2002. In August 2006, they set up the European Commission–Norway Energy Co-operation Group at the senior officials' level, with the task of making concrete proposals to the meetings.[27]

Norway's position as the third largest oil exporter globally after Saudi Arabia and Russia, plus its gas reserves, implies that it is not only their direct competitor, but in the European playing field also shares some interests with the latter. One of them is long-term contracts for gas supply, taken into account the costs for building and maintaining pipelines, and the high investment needs in order to take into use fields located in difficult and environmentally hazardous conditions, especially those in the Barents Sea where commercial production based on expensive LNG technologies is first set to start in the Snøhvit field in 2007. The Norwegian expertise accumulated in Snøhvit and elsewhere prompted Russia to propose a Norwegian–Russian 'strategic partnership' in energy questions. Statoil and Norsk Hydro were correspondingly short-listed by the Russian company Sevmorneftegas (Gazprom and Rosneft's subsidiary), together with two US based and one French company, to develop the giant Shtokman gas field in the Barents Sea (see Figure 6.1). LNG technology had figured for long in the Russian companies' plans for Shtokman's development. But in autumn 2006, the Russian parties made public their decision to instead build a pipeline feeding Shtokman's gas southwards into the Nord Stream pipeline that is to be built from Primorsk to Germany (see Chapter 5). The decision to abandon LNG solutions in the Arctic also paved the way for the announcement that Norwegian and other foreign companies would only be used as sub-contractors in Shtokman. Regardless of this setback, Norsk Hydro continues to be involved in the development of the Kharyga field in the Archangelsk area.

---

25 See e.g. Steensnæs, 'Norwegian Petroleum Policy', speech by the Minister of Petroleum and Energy, at the 14th International Petroleum Tax Conference, 11 November 2003.

26 Matlary, 'The Nordic Countries and EU Membership', p. 241.

27 Ministry of Petroleum and Energy of Norway, 'Terms of Reference of the EC – Norway Energy Cooperation Group', <http://odin.dep.no/oed/english/news/news/026031-990089/dok-bu.html> (accessed 4 September 2006).

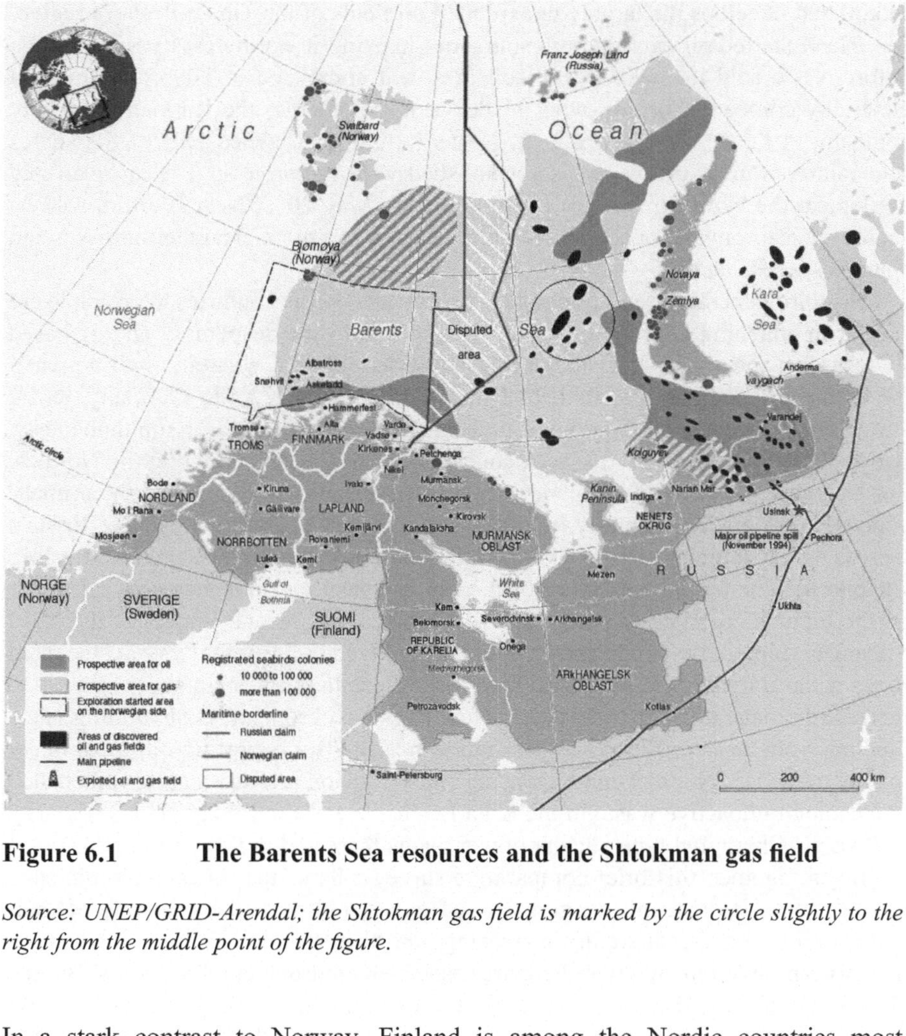

Figure 6.1 The Barents Sea resources and the Shtokman gas field

Source: UNEP/GRID-Arendal; the Shtokman gas field is marked by the circle slightly to the right from the middle point of the figure.

In a stark contrast to Norway, Finland is among the Nordic countries most pronouncedly dependent on energy imports from Russia despite its respectable generation of energy from renewable sources.[28] About 40 per cent of the energy consumed in Finland is imported from Russia in the form of oil, natural gas and coal. The net import of electricity is around 12 per cent of the consumption.[29] The Finnish oil company Neste (formerly part of Fortum) imports oil from Russia to its refinery in Sköldvik and had a joint venture SeverTEK with the Russian company

28 23 per cent of Finnish primary energy consumption came from renewable sources in 2005. Yet, part of Finland's renewable electricity use came from foreign generation via Nordpool; see European Commission, 'Overview'.

29 Nina Wessberg, 'Mitä on suomalainen energiantuotanto?', on-line publications of the Department of Regional Studies and Environmental Policy, University of Tampere (1999), <http://www.uta.fi/laitokset/yhdt/artikkelit/1999_suom_energia.pdf>.

Lukoil that develops the largely unexploited oil fields of the Timan Pechora region. SeverTEK started oil extraction in southern Shapkino in northwest Russia in 2003, before Neste sold its stake to LUKoil in a deal completed in 2005. The Finnish electricity company Fortum owns 21 per cent of Lenergo, the Russian electricity company in the St. Petersburg area. Neste/Fortum also owned half of the shares of a joint venture North Transgas established with Gazprom in 1997 to plan and implement the Nord Stream gas pipeline. But in May 2005, Neste/Fortum sold its shares to Gazprom and hence let the pipeline assume a more clear German–Russian character (see Chapter 5).

The third general observation relates to how the Nordic countries in principle are similar in attaching a strong environmental accent to the deepening, energy based EU–Russian integration. All four boast a political culture wherein environmental issues have become well-established in a similar manner as in Germany, and a strong civil society including environmental lobbying groups. But simultaneously, two regionally differentiated sets of concerns can be discerned among the Nordics. For Denmark, Sweden and Finland, which have a Baltic Sea coastline, the primary concern is the environmentally fragile Baltic Sea, and most pronouncedly, Russia's increasing oil shipments there as a result of new ports around St. Petersburg (see Chapter 8). The partial bypassing of the Baltic transit corridor increases oil traffic on the Finnish coastline in particular, while the environmental impact of the Nord Stream gas pipeline evoked heated comments in Sweden.[30] But for Norway, the main focus is the Barents Sea, where oil shipments close to Norwegian shores are set to increase dramatically with the project of four Russian oil companies of building a new pipeline from western Siberia to Murmansk by 2007 for exporting oil, most likely mainly to the US. Norway and increasingly also Finland are very strongly interested in cleaning radioactive waste in the Kola Peninsula and improving nuclear security in Russia's Chernobyl type nuclear power plants there and in Sosnovyi Bor.

Taken together, this brief comparative survey tells us that in each Nordic state at least some interest in engaging Russia exists, although the channels of their realization differ; that there are increasing, yet divergent energy trade interests in the Russian direction; and that there are associated environmental concerns. But the Norwegian and Finnish cases represent the extreme ends among the Nordics. As immediate neighbours they have a special need to engage Russia. In energy trade they are mirror images of each other, with Norway being a major energy exporter, and Finland importing on a large scale from Russia. And in environmental terms, Norway is primarily interested in the Barents Sea, while Finland looks at the Baltic

30 In autumn 2006 PM Göran Persson threatened to stop the project if the risks for the environment would be too high. The Swedish Minister of Environment Lena Sommerstad claimed that Sweden decides whether or not the pipeline is built. The pipeline passes trough the Swedish economic zone in the Baltic and the pipeline route goes near nature conservation area in Gotland islands; See Juhani Roiha, 'Ruotsi vauhdittaa Itämeri-yhteistyötä – Kiusanaiheena Venäjän ja Saksan yhteinen maakaasuputki,' *Turun Sanomat*, 17 August 2006; Maynie Rantala, 'Itämeren kaasuputkea vastustetaan Ruotsissa', *Sveriges Radio Sisuradio*, 8 August 2006, <http://www.sr.se/cgi-bin/sisuradio/nyheter/artikel.asp?artikel=912737>. Sweden's position on Nord Stream pipeline issue is important, because gas pipelines need pumping stations to be built along the route and Gotland Island's location is pivotal here.

Sea. In other words, looking more closely at the narratives put forth by Finnish and Norwegian actors give us an idea of the continuum of Nordic ideas vis-à-vis EU–Russia energy relations, and of what problems and prospects there are for a dialogue.

Norway and EU–Russia Energy Dialogue

From Detachment to Engagement via Energy Trade

In July 1992, Norway and Russia established a Forum on Energy and Environment, which functions as a permanent co-operation body for the energy ministers. The working groups cover energy legislation, energy policies, environmental issues, industrial issues, hydro power and energy conservation. In addition, the Norwegian and Russian governments can exchange views on the subject in the context of Barents co-operation, the ND, and the EU–Russia energy dialogue.[31] Overall, the history of the Norwegian government's engagement of Russia is relatively short, but at the same time it is a story of continuous expansion since the early 1990s, despite some remaining historical suspicions. In 1998, in a speech in Washington, the Norwegian State Secretary Åslaug Haga mentions that although the Red Army liberated the eastern part of the Norwegian county of Finnmark in the north of the country in 1944–1945, 'mental distance' between the two countries grew quickly again and strengthened even more during the Cold War. The words ending his presentation are symptomatic of the historical legacy: 'co-operation with the great Russian bear has never been easy, and might never be, but it is even more important today, than ever.'[32]

The perceived difficulties of engaging the 'Russian bear' initially led to embedding the bulk of mutual relations firmly into multilateral contexts, although for example the BEAC co-operation was initially launched with the help of bilateral consultations.[33] Since then, the multilateral element has kept Barents co-operation going,[34] despite the fact that it has become largely immersed into the EU context where Norway lacks voting rights. But apart from the Barents and EU/ND contexts, another particularly pivotal external determinant is the broader transatlantic relationship which traditionally has in the politico-strategic sense been more important to Norway than the EU direction which has connoted mainly markets. State Secretary Haga makes it clear that Norway's engagement of Russia takes place firmly on the basis of western values and principles:

> In a concerted effort together with our neighbours and partners on both sides of the Atlantic, we are applying the instruments available to us to achieve a common goal: a

31 Skjelbred, 'Opportunities in the Russian Oil and Gas Offshore Market'.

32 Åslaug Haga, 'Norway and Russia – a Northern Connection', speech in the Carnegie Endowment for Peace, Washington, 19 October 1998.

33 Johan Eriksson, 'Security in the Barents Region: Interpretations and Implications of the Norwegian Barents Initiative', *Cooperation and Conflict*, 30/3: 279.

34 Haga, 'Norway and Russia'.

> Europe in peaceful co-operation from the Atlantic to the Urals. *Our objective is a larger community of nations patterned on the last fifty years of co-operation west of the former Iron Curtain.* It is *within this context* that Norway is seeking to play a particularly active role in relation to its neighbouring areas, upon which we are dependent, and which in turn are dependent on us.[35]

However, regardless of Haga's references to transatlantic values as the basis of engaging Russia, more recently there have been signs of increasing pragmatism. This is also evinced in the progress made in the delimitation of the Norwegian–Russian maritime border, an issue that is also linked to the disputes on fishing rights within the shady Arctic frontiers.[36] In energy trade issues, achieving transparent and EU single market compatible legislation is what the Norwegian Government most pronouncedly is after. Norway's partly state-owned companies are interested in accessing the Russian market, but what they see as Russia's inadequate legislative framework has hindered the realization of these interests.[37] Lack of investments into infrastructure and into the economy of northwest Russia from the Norwegian side is what at least occasionally has made the Russian party frustrated.[38] But on balance, it must be noted that in the Shtokman case where the Norwegian companies had already invested a lot of efforts during the bidding process, the Russian party's decision in autumn 2006 to rule out stakes for foreign companies hindered investment possibilities seriously.

In its 2006 budget, the Norwegian Government provided 130 million Norwegian kroons to bolster its efforts in the high north, including co-operation with Russia. However, most of these funds went to strongly Norwegian concerns like geo-scientific research with a strong environmental tone, nuclear safety in the Kola peninsula, and for strengthening the Barents secretariat in Kirkenes.[39] A further 110 million kroons were planned to be allocated for nuclear and radiation safety in Russia, and also a small sum to the restoration of the Russian section of the highway leading from Kirkenes to Murmansk, with the Murmansk region providing the lion's share itself.[40]

---

35 Haga, 'Norway and Russia'; emphasis added.

36 RIA Novosti, 'Russia, Norway Make Progress in Border Talks', 17 February 2006 <http://www.barentsobserver.com/index.php?id=306490&cat=16282&xforceredir=1&noredir=1 (accessed 22 February 2006).

37 Haga, 'Norway and Russia'; Skjelbred, 'Opportunities in the Russian Oil and Gas Offshore Market'.

38 E.g. Hønneland, 'Identity Formation in the Barents Euro-Arctic Region', 293.

39 Ministry of Petroleum and Energy of Norway, 'Continued focus on High North Issues', <http://www.dep.no/odinarkiv/english/bondevik_II/oed/026031-070398/dok-bn.html> (accessed 24 February 2006). In early 2007, some domestic observers started arguing that Norway erroneously continues to support environmental and other projects in Russia, with too little or even contrary results, in a situation where economically strengthened Russia could also itself make similar investments; e.g. Terhi Width, 'Arvostelijat: Norjan Venäjä-politiikka jämähtänyt paikalleen', *Helsingin Sanomat*, 3 March 2007.

40 Bellona Foundation, 'Norway to Allocate 110m Crowns for Russian Radiation Security', <http://www.bellona.no/en/international/russia/navy/co-operation/42042.html> (accessed 22 February 2006); Ria Novosti, 'Norway Allocates $145,500 to restore road

Norway in the EU–Russia Energy Dialogue

Norway's accent on market principles for investing into Russia echoes the EU's markets and competition principle and its consequent message to Russia in the context of the energy dialogue (see Chapter 1). But it also represents a wider Norwegian strategy of opening the country's own energy sector for foreign investment. Norway initially opened its continental shelf on a free markets basis, but since the larger discoveries in the 1970s, state involvement and centralized optimization of the whole work programme became the key words, with a 50 per cent state participation becoming a rule for licensing (including both state institutions and state-bound companies).[41] The maturing of the Norwegian continental shelf has highlighted the need for different types of expertise available outside the country.[42] A related Norwegian interest lies in expanding the global market for Norwegian oil. In addition to pipeline supplies of gas to the European market, in the future, there will be a growing interest in LNG that can more easily be re-directed than pipeline supplies.

To promote Norwegian energy sales, the Norwegian Government established INTSOK (Norwegian Oil and Gas Partners) in 1997. Norway also has OG21 (National Technology Strategy) for supporting the country's energy sector service industry. In 2004, the Norwegian Government proposed to increase its funding for petroleum research by 80 per cent to about 300 million kroons in the context of the DEMO, PETROMAKS and other research programmes including both short and long-term research. As much as 80 per cent of the world's subsea equipment market is claimed to be served by Norwegian-linked high-technology companies.[43]

To protect market access in general, Norway participates in the IEA under a special association agreement, and takes part in the WTO. But in the Norwegians' discussions in the EU direction, it is interesting to note how alongside the market principle suddenly emerges the security of supplies principle.[44] It is not often found in Norway's internal debate as a result of its meaninglessness in a country with abundant reserves of its own. But in this connection the Norwegian Minister of Petroleum and Energy concomitantly promotes the country's market considerations by proposing that CO_2 captured in Europe would be injected into North Sea oil fields in order to enhance oil recovery, a technology developed in Norway. In the narrower Nordic context, a slightly different mention of the security of supplies principle is

---

in Russia's Arctic', <http://en.rian.ru/world/20060215/43568872-print.html> (accessed 20 February 2006).

41 Farouk Al-Kasim, *Managing Petroleum Resources: The 'Norwegian Model' in a Broad Perspective* (Oxford: Oxford Institute for Energy Studies, 2006), 144–5; 242.

42 Thorhild Widvey, 'European Energy – Working without Borders', speech by the Minister for Petroleum at the ONS Stavanger, 25 August 2004.

43 Widvey, 'Priorities in Supporting the Oil and Gas Industry Abroad', speech by the Minister for Petroleum at the annual conference on the internationalization of the oil and gas industry, Stavanger, 9 November 2004; see also Al-Kasim, *Managing Petroleum Resources*, p. 244.

44 Ministry of Petroleum and Energy of Norway, 'The Energy Dialogue between Norway and the European Union', <http://www.dep.no/odinarkiv/english/bondevik_II/oed/026001-070234/dok-bn.html> (accessed 24 February 2006).

found, referring to Norway's seasonal dependence on electricity supplies from the Nordpool market.[45] This, however, can be addressed through building new gas-fired power plants.

Due to its own energy dialogue with the EU and a separate track of bilateral dialogue with Russia, Norway can offer a lot to the EU–Russia energy dialogue. Thanks to its resources and supplies, and its relatively solid commitment to the EU's markets and competition principle, Norway can be taken as the Union's ally. But as a result of its own access to Russia and the emerging partnership between Norwegian and Russian energy actors, despite the setbacks in Shtokman's case, Norway simultaneously remains aloof the EU side. One might further comment that the EU's security of supplies principle is important to Norway chiefly to the extent that it helps to maintain Norway's energy markets within the EU. Here Norway is in the same camp as Russia as a major energy supplier, although from the Union's point of view, their existence as separate suppliers helps to diversify energy imports.

The Environmental Limits to Engagement and Trade

Norway's EEA membership has led to the adoption of stricter environmental standards in some key fields, and to Norwegian participation in the diplomatic efforts that led to Russia's adoption of the Kyoto protocol.[46] But although Norway has gradually approximated the EU's sustainability policies, it has its own and distinctive environmental agenda. For example, since the early 1990s, before giving its backing to the Kyoto protocol, Norway has operated an internal CO_2 tax.[47] Energy exploitation and environmental matters are closely connected in strikingly many statements by Norwegian policy-makers, ranging from high officials in the energy field to environmental policy, the Prime Minister's office, and energy industry:[48]

> Coexistence between environment, fisheries and petroleum industry is our goal, and we need new technology to solve the environmental challenges. In order to achieve this we must involve industry and other stakeholders.[49]

---

45 Widvey, 'Nordic Energy Policy', speech at the Nordic Council, Oslo, 10 December 2004 <http://www.dep.no/odinarkiv/english/bondevik_II/oed/026021-090110/dok-bn.html> (accessed 24 February 2006).

46 See also Kim Traavik, 'How Does EU Policy Affect a Non-Member State?', speech in the Conference of Peripheral Maritime Regions, Stavanger, 24 September 2004.

47 The Norwegian Oil Industry Association (OLF), however, termed the CO_2 tax a mere money-spinner for public coffins and ineffective for environmental purposes after the start of the Kyoto stipulated emissions trade. As a replacement the industry offered to purchase Norway's Kyoto stipulated emissions quotas from which it had previously been exempted due to the CO_2 tax and the government paying for the national emissions; see The Norwegian Oil Industry Association (OLF), 'Wants to help Norway Meet Climate Commitments', <http://www.olf.no/english/news/?28816> (accessed 22 February 2006).

48 See also Al-Kasim, *Managing Petroleum Resources*, p. 244.

49 André Støylen, 'Expectations of the Norwegian Government on Environmental Reporting', speech by the State Secretary, Ministry of the Environment, Workshop on Oil & Gas Sector Environmental Reporting, Stavanger, 8 November 2002.

> Norway gives high priority to developing our relations with our Russian neighbour in a wide range of areas. The Barents region has tremendous energy potential. The decisions that are made about exploration, development and transport in this area will have important consequences for the future. The Barents Sea holds some of the world's richest fisheries resources. But these Arctic areas are environmentally extremely sensitive. Therefore, it is of the utmost importance that petroleum exploration in the High North fully reflects the special needs and requirements of the Barents region.[50]

> Environmental considerations are an integral part of the oil industry's framework conditions. Together with the industry's efforts to develop measures and creative environment-friendly solutions, this has led to the Norwegian petroleum sector leading the way in this area. This positive environment trend is illustrated not least in the increased focus on energy efficiency and *the work towards zero environmentally harmful discharges to the sea by the end of 2005*.[51]

The chorus of Norwegian actors linking energy exploitation and transportation with environmental and fisheries issues is amplified by the Government's adoption of an integrated management plan for the Barents Sea in March 2006. Several assessments were carried out in order to comprehend the interconnections between oil drilling, transportation, fisheries industry and the environment. As a result, a map was drawn that outlines extensive areas where oil extraction activities would not be started during the Government's term.[52] The Government is also committed to co-operating with the industry in order to capture CO_2 in all gas fired power plants by 2009.[53] However, the Norwegian Government's and industry's efforts notwithstanding, it is clear that sustainability policies are not set in stone and that critical environmental debate in a typical Nordic fashion will continue. For example, in August 2005 Friends of the Earth Norway found out that oil companies had flushed more than 80 tons of the toxic pollutant Perfluorooctane sulphonate (PFOS) from Norwegian oil rigs directly into the North Sea. This led Statoil and Norsk Hydro to announce in October a change in their practices in the next few months.[54]

---

50 Kjell Magne Bondevik, 'Strategic Perspectives on Reliable Energy Resources in Europe: Norwegian–German cooperation', Prime Minister's speech at industrial seminar, Government Guest House, Oslo, 28 September 2004.

51 The Norwegian Oil Industry Association, 'Global Industry with Regional Significance: Social Perspectives', p. 28; emphasis added.

52 See Government of Norway, 'Outline of an Integrated Management Plan for the Barents Sea' <http://odin.dep.no/filarkiv/237456/Outline_of_an_Integrated_Management_Plan_for_the_Barents_Sea.pdf>; The Office of the Prime Minister, 'Integrated Management Plan Ready', <http://www.dep.no/smk/english/news/press_centre/press_releases/001001-071466/dok-bn.html> (accessed 5 September 2006).

53 Helen Bjørnøy, 'Minister of the Environment Speech in Montreal', eleventh session of the Conference of the Parties to the Climate Change Convention and first meeting of the Parties to the Kyoto Protocol, Montreal, 7 December 2005 <http://www.dep.no/md/english/news/speeches/022001-090136/dok-bn.html> (accessed 23 February 2006).

54 Friends of the Earth Norway, 'Norway: Exposed Oil Companies To Stop Using Toxic Pollutant', <http://www.naturvern.no/cgi-bin/naturvern/imaker?id=71734> (accessed 22 February 2006).

There are some indications that Norway's similar status with Russia as a major energy exporter, and its emerging partnership with Russia, can contribute significantly to the incorporation of sustainability into energy projects realized in northern Europe. This should be in the EU's interest. Time and again the Norwegians stress the environmental frame conditions for any economic co-operation in Arctic areas. During the intensive negotiations for a Norwegian–Russian deal for the Shtokman project in spring 2006, the Norwegian Foreign Minister offered using Norway's experience for drafting a joint, environmentally sensitive management plan for the whole Barents Sea. In a speech in Moscow, he maintained that Norway and Russia 'need to agree on applying the best available technology and most stringent environmental standards'; that will enable to 'fulfil the responsibility we have as coastal states in a region with a vulnerable environment'.[55] Although the Russian party's decision to involve Norwegian companies only as sub-contractors in Shtokman was a serious disappointment, by no means does it preclude entirely their future influence on the incorporation of the sustainability principle. The Kirkenes declaration in July 2005 by Norwegian, Russian, US and EU political and business decision-makers already indicated how with Norwegian lead, environmental issues can be incorporated into Arctic energy policy:

> Developing the energy potential of the north, while ensuring strict environmental and safety standards and compliance, is therefore a joint task for governments and companies involved. This will also form the basis of realising the value creation potential within the framework of sustainable development.[56]

Finland and EU–Russia Energy Dialogue

The Story of Engaging Russia

Since the beginning of the 1990s, the dominant policy narrative directing Finnish policies towards Russia asserts that Russia needs assistance in order to combat threats, mainly environmental, impacting Finland across the border. In 1992 this policy was named neighbouring area co-operation of the Government of Finland. An interview with a Finnish official from the Ministry of Environment illustrates this narrative:

> It was quite a shock in Finland when the state of the environment (in the Soviet Union) was revealed. There were a lot of horror stories… and that is why we took the environment as the priority and began the neighbouring area co-operation in the beginning of the 1990s. We began by preparing a report in which we identified the possible targets of co-operation. (…) And on the basis of the report we built a process and created connections already during the reporting. (…) It was a conscious choice that

55 Jonas Gahr Støre, 'Norway – A Cooperation Partner in the North', speech in the MGU (Moscow State University), Moscow, 17 February 2006, <http://www.odin.dep.no/ud/english/news/speeches/minister_a/032171-090516/dok-bn.html> (accessed 21 February 2006).

56 'The Arctic Energy Agenda', <http://odin.dep.no/filarkiv/252695/Artic_Energy_Agenda_Felleserklaring_Kirkenes2.pdf> (accessed 20 February 2006).

we did not start only on the governmental level but took the bottom-up approach that underlined the role of industry and commerce.[57]

From the total budget of neighbouring area co-operation policy with Russia, about one third is directed to the environmental sector, making up approximately 30 million euros during 1991–2001.[58] Protection of the Baltic Sea is a top priority. Co-operation in nuclear safety issues with Russia involves annually approximately two million euros.[59] On the basis of an evaluation of threats, assistance is primarily addressed to the improvement of the safety of nuclear power plants in Finland's vicinity in Sosnovyi Bor and the Kola peninsula.[60] But co-operation in nuclear safety can also be read as an effort to legitimize the Finnish imports of electricity produced in these power plants. When the construction of a fifth nuclear reactor in Finland was discussed in the Finnish Parliament in 2002, safety standards in Russian power plants were often mentioned as being poor. One argument for the construction was that 'it is irresponsible to prolong the lifetime of old reactors by increasing the purchase of electricity from Russia'.[61]

In the mid–1990s, the narrative guiding Finnish policies towards Russia assumed a by-plot: since Finland's bilateral aid can only be minimal compared to the needs, multilateral co-operation is needed.[62] The ND initiative was pivotal for this end, and represents a wider Finnish narrative on how the EU should engage Russia. This narrative emphasizes interdependence between the Union and Russia. This is most evident in the energy sector:

> The northern dimension of the European Union is best crystallised in energy co-operation of the beginning of the next millennium. Europe needs energy, and Russia markets for its energy exports. The interdependency of Europe and Russia is a fact and an important basis of new co-operation.[63]

It is hoped that in the future, the economic potential of Russia's natural resources and the energy sector will solve the 'environmental game' that is the primary concern

57 Interview 1, Ministry of the Environment 2002.

58 Interview 2, Ministry of the Environment 2002.

59 Interview 5, Ministry for Foreign Affairs 2005.

60 Säteilyturvakeskus, *Säteilyturvakeskuksen raportti: Säteilyturvallisuutta yli rajojen, Säteilyturvakeskuksen lähialueyhteistyö 1992–2002* (Helsinki: Säteilyturvakeskus, 2002).

61 Kalervo Kummola (MP, National Coalition Party), speech in the Finnish Parliament, 23 May 2002.

62 E.g. Ministry for Foreign Affairs, 'Finland's Strategy for Cooperation in the Neighbouring Areas', Department for Russia, Eastern Europe and Central Asia, Helsinki, 1996; 'Finland's Strategy for Cooperation in the Neighbouring Areas', Department for Russia, Eastern Europe and Central Asia, Helsinki, 2000. Note that early in the new millennium, over half of the Finnish finance in the environmental sector, for instance, went to multilateral co-operation and to the activities of international financial institutions such as EBRD, Nordic Investment Bank and Nordic Environmental Financing Company (Nefco); see interview 2, Ministry of the Environment 2002.

63 Martti Ahtisaari, 'Europe Needs Russia and Russia Europe', speech in Kremlin, Moscow, 27 November 1997.

of Finland in its relations with Russia.[64] It is also understood that other EU members that are geographically distant from Russia are less interested in the threats Russia poses than in the possibilities it offers. Besides, Russia itself is keener on energy than environmental issues.[65] Thus, it is easier to 'sell' the idea of mutual co-operation to the EU and Russia by stressing the positives that unites them, i.e. energy issues. This has made Finnish non-governmental organizations, among others, concerned about the utilitarian nature of the ND and other co-operative instruments of the EU that Finland promotes. For example the Secretary-General of the Finnish–Russian Society Merja Hannus presented her concern in a national ND conference in Oulu:

> Activities of the northern dimension have to bring benefits to people and the environment. The horror scenario about the northern dimension, in which the dimension is merely a project aimed at the utilisation of natural resources of Russia, is not supported by us anyone. (...) However, NGOs want to remind that the principles of public participation and sustainable development need to be carried out in the policies of the northern dimension at all levels and in all projects.[66]

Governmental actors alike demand the integration of environmental issues in all sectors of EU–Russia co-operation, especially the energy dialogue. The importance of the question is evident in the growing oil transportations in the Gulf of Finland: between 1995–2005 the volume of transportation grew from 20 million tonnes per year to 100 million tonnes.[67] This development gave rise to concerns about the safety of new Russian terminals and transports. When the Russian oil pipeline operator Transneft was constructing the oil terminal in Primorsk (Koivisto), Finnish governmental and non-governmental actors pressed heavily the Russian counterpart in order to prompt them to conduct an environmental impact assessment of the project. The pressure brought results, as a civil servant working for the Finnish Ministry of the Environment describes:

> Well, concerning Koivisto I have to say that for all the Finns who have followed the procedure it has been a surprise how much Russia has actually invested in environmental security of the port. I think the foreign interest has had an impact here, and Russia has understood that they have a very good opportunity to make an impression.[68]

Finland, Estonia and Russia established a mandatory ship reporting system for all vessel traffic in the Gulf of Finland (GOFREP). This system started operating on 1 July 2004. Yet, oil transportation has caused grudge between the Finnish and Russian parties. Russia was reluctant to accept the designation of the Baltic Sea as a particularly sensitive sea area by the International Maritime Organisation (IMO). The

---

64 Interview 3, Ministry of the Environment, 2002.

65 Interview 5, Ministry for Foreign Affairs, 2005.

66 Merja Hannus, 'The Northern Dimension – the Aims and Activities of NGOs', speech in the national forum of the Northern Dimension, Oulu, 15 January 2001.

67 Saara Hänninen and Jorma Rytkönen, *Oil Transportation and Terminal Development in the Gulf of Finland* (Helsinki: VTT Publications, 2004); Finnish Ministry of the Environment, Communication, 6 April 2005.

68 Interview 4, Ministry of the Environment, 2002.

Baltic Sea, with the exception of Russia's territorial waters, got the status protecting it from increased shipping and illegal oil dumping in 2004.[69] Before the designation, Finland very actively lobbied the EU Commission so that it would put pressure on Russia in this issue. The Finnish Minister of the Environment, Jouni Backman, and his Swedish colleague, Lena Sommestad issued an appeal in the Swedish newspaper *Dagens Nyheter* in February 2003 to their colleagues around the Baltic Sea for improving safety at seas in the region:

> (…) The catastrophe of the oil tanker Prestige last autumn aroused discussion in the Nordic countries about the effects of a similar accident in the Baltic Sea. During the last weeks and days the concern has been directed to the danger caused to the whole Baltic Sea by oil transports coming from Primorsk in the eastern Gulf of Finland. The Greek oil tankers Stemnitsa and Minerva Nounou, which brought oil from Primorsk, do not meet the requirements of ice classifications. The Baltic marine environment commission (HELCOM) is already working for the better marine safety in the Baltic Sea, but additional measures are necessary. Heavy oils should not be transported in single-hull vessels. We also need standardised and strict rules for the ice classification of vessels working in the Baltic. Winter traffic should be limited to vessels equipped to cope with pack ice conditions and that comply with Nordic specifications for ice classification. More effective piloting and towing schemes would also improve marine safety. The ministers of the environment of Finland and Sweden appeal to their colleagues around the Baltic Sea and to the European Union to work towards improved marine safety in the Baltic Sea within HELCOM and other relevant organisations.[70]

The increased oil transport did not remain solely a governmental worry in Finland. World Wildlife Finland and another Finnish environmental NGO Luonto-liitto have their own anti-oil leakage troops to assist in possible oil catastrophes. But all this concern related to Russia prompted the Finnish MP Kimmo Sasi to warn that there is an apparent risk of a split between the eastern part of the Baltic Sea region with its problems of nuclear safety, the environment, and waste water, among others, and the western part of the region with its positive development concerning infrastructure projects, transport corridors and bridges. Accordingly, in his opinion, in the region's eastern parts threats dominate, whereas in its western part, it is possibilities.

Even though Finnish governmental actors like to underline the EU–Russia interdependency in energy issues for the European audience, the speech of the Finnish Minister for Foreign Affairs Erkki Tuomioja indicates that the environment is always a key element for Finland also in the context of EU–Russia relations:

> As to our Finnish interests in this (EU–Russia cooperation) framework, environmental issues are of major significance. As bordering neighbours, the environment is clearly our common concern, ranging from the climate change to nuclear safety, safety of maritime transport and the state of our border waterways.[71]

---

69 Hänninen and Rytkönen, *Oil Transportation and Terminal Development*, 23.

70 Jouni Backman and Lena Sommestad, *Dagens Nyheter*, 18 February 2003.

71 Erkki Tuomioja, 'Baltic Sea Cooperation at a Crossroads', speech at the 10th Anniversary Seminar of the Baltic Institute of Finland, Tampere, 20 October 2004.

Finland in the Energy Dialogue

The strong emphasis on environmental protection prevails also in the bilateral energy sector co-operation between Finland and Russia. However, energy was introduced as a specific focus area into the Finnish neighbouring area co-operation only in the 2000 strategy of the Ministry of Foreign Affairs.[72] During 1991–2001, Finland supported the non-nuclear energy sector of Russia by 2.681 million euros in total, which is much less than for example Denmark, which put up 5 times more funds on the Russian energy sector, but less than one tenth of the Finnish funding for the environmental sector. According to an evaluation of energy sector co-operation with CEE countries during 1991–2001, Finland funded mainly projects dealing with district heating and energy efficiency in Russia. Finland also supported the use of renewable energy sources and the enhancement of energy efficiency especially in northwest Russia.[73] Now that the Kyoto Protocol has entered into force, energy sector co-operation will have an even more environmental overtone. In spring 2005, the relevant ministers of Finland and Russia agreed on co-operation related to the protocol, for example on the development of joint implementation projects between Finland and Russia.[74] Thus, it seems quite evident that also in the future the primary emphasis in co-operation with Russia will be on the environment rather than on energy. This is admitted in the Ministry of Trade and Industry:

> I suppose that the co-operation will be strongest in sectors where we undergo definite threats. Environmental co-operation probably keeps it leading post, and nuclear safety is important. There exists a broad consensus in Finland that it must be continued because of the security of our citizens.[75]

Finland agrees with the strategic level objectives of the EU–Russia energy dialogue and admits that they may benefit also Finland. Within the dialogue, Finland has shown a major interest in the diversification of the energy mix and the development of new transport routes particularly for gas, underlining the country's location in a junction of energy transport routes. Finnish actors supported the construction of the Nord Stream gas pipeline with the argument that it would be of strategic importance not only to the region but to the whole of Europe, particularly because it would improve security of supplies by providing a new and independent route to Germany and prospectively further downstream. But in Finland, diversification of supplies to the Norwegian direction also finds support.

These points related to the diversification of supplies demonstrate, however, that in general governmental actors, as well as the public, prefer lessening the dependency on Russian energy on the national level. The debate on the decision of constructing

---

72 Ministry for Foreign Affairs of Finland, 'Evaluation of Energy Sector Co-operation Financed by the Government of Finland as Part of the Action Plan for Central and Eastern Europe', Helsinki, 2002.

73 For further details, see Ministry for Foreign Affairs of Finland, 'Evaluation of Energy Sector Co-operation', p. 49.

74 Finnish Ministry of the Environment, Communication, 7 June 2005.

75 Interview 6, Ministry for Trade and Industry, 2005.

the fifth nuclear reactor in Finland clearly shows, too, that in Finland Russia is seen rather as a threat than a possibility, or normal partner, in energy issues. A deputy of the Centre Party ironically noted in the parliamentary discussion that if the case really is, as many other deputies in the discussion had repeated, that natural gas is not an option and that the import of gas is a risky business, then 'we, to be honest, should tell this to other EU countries which import relatively more gas than we do'.[76]

Businesses do not equally demonize the dependence on Russian energy. What is peculiar to Finland is that even though businesses, naturally, welcome the EU–Russia energy dialogue, they have quite neutral attitudes towards it. The speech in Oulu in 2001 by Jaakko Ihamuotila, the CEO of Fortum at that time, crystallizes the attitudes of the Finnish business elite:

> From the Finnish point of view the decision (about the establishment of the EU–Russia energy dialogue) must be considered as a welcomed decision, even though a rather late one. Finnish energy companies have acted and made investments in Russia for over ten years already. In fact the whole period of bilateral trade reflected the same idea: we need to import in energy so why not to import it from our neighbour, and in exchange increase our export there.[77]

Finnish Expertise in Russian Affairs

The policy narrative by which Finnish actors justify the Finnish position in the EU–Russia co-operation asserts that because of the history and the geographical proximity of Finland to Russia, Finland is a specialist in Russian affairs. This is affirmed by repeating that Finland has a long experience of co-operation with Russia and its predecessors. Thus, other countries and the EU can follow the Finnish model of good co-operation with Russia, as President Tarja Halonen suggested in Berlin in 2000:

> In addition to the regular dialogue of the political leadership we have condensed co-operation with the Russian regions adjacent to Finland. Earlier even very tiny practical problems had to be handled through Moscow. Now we can, on the basis of the co-operation agreement of 1992, enforce environmental, educational and other co-operation directly with Russian regions. This is a significant improvement. At the same time we have established connections between actors that solve practical problems. This model can be applied also in the co-operation between Russia and its other neighbours.[78]

This experience and the position as a specialist in Russian affairs can and should be taken advantage of in the EU, as President Halonen notes in the interview published in the magazine of the Confederation of Finnish Industries in winter

---

76 Hannu Takkula (MP, Centre Party of Finland), speech in the Finnish Parliament, 23 May 2002.

77 Note that some of the Finnish investments to Russia claimed by Ihamuotila have since then been sold back to Russian companies; see above and Jaakko Ihamuotila, 'The Northern Dimension and Energy', speech in the National Forum of the Northern Dimension, Oulu, 15 January 2001.

78 Tarja Halonen, speech in the 'Quandt Stiftung' seminar, Berlin, 24 November 2000.

2005.[79] Halonen reminds that within the EU, Finland cannot play individual games with Russia – the palette of possible measures in any case stems from the strategic level relations between the EU and Russia. Nevertheless, a Finnish official argues that Finland is a much better advocate of Russia, on the one hand, than 'a country for which Russia is a theoretical definition somewhere very far away', and, on the other hand, than for example the Baltic States where attitudes towards Russia are much more negative.[80]

At the same time, it is widely complained in Finland that the alleged Finnish expertise is not appreciated enough within the dialogue between the EU and Russia. Moscow often seems to be more interested in working with large EU countries, especially Germany. The EU Commission likewise fails to appreciate Finnish expertise:

> In Russia the Finnish expertise is somehow appreciated. I have not noticed the same outside Russia. For example now that I work in energy working groups, I have made no secret of my ten-year experience in the co-operation between Finland and Comecon, but…[81]

One example indicating the low level of appreciation of the Finnish expertise on Russia is considered to be the slow progress within the ND process, the northern European testing ground for the energy dialogue (see Chapter 4). A former member of the European Parliament, now in the Finnish Parliament, mentioned that Finland should not wait that 'somebody in Europe says that it is a great thing to have this economic zone of Northern Europe or the northern dimension'.[82] The EU Commission, perceived as inactive, receives the sharpest critique from Finnish actors especially in energy issues. It has been said, among other things, that because of the Commission, the dialogue has hardly evolved at all. The Finnish Prime Minister Paavo Lipponen rumbled on the Commission already in 2001:

> Co-operation on energy cannot be based merely on the calculations of the price of oil. It has to be persistent and profitable to both counterparts. You can call to Moscow also when the oil prize does not touch the clouds.[83]

The main problem allegedly is that the Commission has not consulted member countries but has tried to negotiate directly with the Russians. This has not led to any significant results, because the Commission does not have a full mandate in energy

---

79 The Confederation of Finnish Industries, 'Suomen asemaa Venäjän suhteissa turha kainostella', *Prima*, 1–2 (2005).

80 Interview 6, Ministry for Trade and Industry, 2005.

81 Interview 6, Ministry for Trade and Industry, 2005; Comecon, or the Council for Mutual Economic Assistance, CMEA, denotes the economic organisation that from 1949 to 1991 linked the USSR with the countries of Eastern Europe.

82 Marjo Matikainen-Kallström (MP, National Coalition Party), speech in the Finnish Parliament, 17 November 2004.

83 Paavo Lipponen, speech in the national forum of the Northern Dimension, Oulu, 15 January 2001.

issues.[84] Also the Russian side has caused problems, especially the turbulence within its administration. It has not always been easy to find out who is responsible for what and who can make decisions.[85]

Finnish actors recommend that the EU would make more pronounced use of a regional approach in its energy dialogue with Russia. Finnish civil servants responsible for the implementation of energy co-operation stress the role of the regional organization BASREC in the realization of the energy dialogue:

> The BASREC as an organisation does not have such legitimacy problems (as the Commission has). Besides, in energy issues it is important that member countries are all properly united, because energy is under common jurisdiction. (…) The issues would be the same as we have advocated in our own policies. Just a bigger machine and with a certain kind of programmatic orientation… Everybody could even work bilaterally... But then the Commission worked as a conductor and would not desperately try to play in a row most loudly. At the moment the commission tries to conduct and runs to the row to play, and then back to conduct… meanwhile the orchestra has disappeared.[86]

Also other practices familiar from the Finnish bilateral neighbouring area co-operation, like operating on a low profile and in co-operation with businesses and the grassroots level, are preferred by Finnish actors. The share of public funding in the implementation of projects under the ND, especially in the energy sector, is inevitably limited. Therefore private companies are needed, as are also smaller scale projects.

That Finland took the development of the ND as one of the top priorities during its EU presidency in the latter half of 2006, explains why the ND, regional co-operation and EU–Russia relations in general were under very lively discussion in Finland prior to assuming the presidency. Three factors were emphasized in the ND policy: first, it is a policy for all member countries, not a regional initiative. Secondly, Russia should be engaged in the policy as an equal partner. Thirdly, Finland should take a more active role, and serve as an example also by increasing the budget of neighbouring area co-operation. The Council of EU ministers within the Finnish Government lined up the Finnish starting points for the further development of the ND in autumn 2004 by deeming the status of the northern dimension in the EU as not strong enough, and this is where the Commission has not filled its responsibility as a leader of the policy. Thus, in the Finnish debate, there were expectations for the country to act, and others were also expected to look for the same. The NDEP partnership and other comparable future projects, for example the one agreed in the transport and logistics sector, represent the Finnish answer for the identified shortcomings of the ND policy so far. An ND energy partnership may be a on the agenda in the future.

---

84 Interview N, Ministry for Foreign Affairs, 2005.

85 Interview 3, Ministry of the Environment, 2002.

86 Interview 6, Ministry for Trade and Industry, 2005.

Conclusion

Of the two Nordic case studies more closely examined in this chapter, Norway is a major energy exporter. Norway's resources, its state bound companies and the mix of competitive and partnership ties they are developing with the Russian parties makes Norway a much more significant actor than mere membership in the EEA and the resulting obligations for conforming with the EU's internal energy market regulation would suggest. Along the markets and competition principle, Norway shares many of the interests and problems that the EU is experiencing vis-à-vis Russia, but concomitantly the separate Norwegian–Russian energy dialogue implies Norway's special status and the possibility of Norway overtaking Finland's traditional role as Russia's number one trusted partner in the north. The EU's security of supplies concerns are meaningless within this context, although Norwegian actors may mention the term whilst discussing with the Union in the context of the EU–Norway energy dialogue. With regard to the sustainability principle, Norway's distinct and relatively successful focus on the Barents Sea can make an important contribution to the more general realization of the principle in northern European energy policy.

Finland differs drastically from Norway in the way security of supplies questions are relevant there in the same way as in most EU members of today. But Finnish actors are not satisfied with the current state of the EU–Russia energy dialogue nor with the EU's overall engagement of Russia. Many Finnish actors, especially those working for energy issues on the governmental level, doubt whether there exists a real energy dialogue between the Union and Russia at all. Finnish actors emphasized low profile methods familiar from the bilateral Finnish neighbouring area co-operation and the ND – i.e., regional approach, small or medium size projects, public-private partnerships – rather than the role of the EU Commission.

It can be argued that up to a certain level, the interests of Finnish actors in the energy dialogue remain amorphous. Finland clearly is not a frontbencher in the EU–Russia energy dialogue, no matter the very real need to import energy or noble words of Finnish statesmen praising Russian energy in Europe. On the political level, the desire is to be rather less than more dependent on Russia in terms of energy imports as evinced in the debate on the construction of the fifth nuclear reactor and the emphasis on diversification of supplies. Yet, most Finnish energy companies conduct their business-as-usual with their Russian partners independently from and without any special interest in the political dialogue, although the controversies in the Swedish–Finnish–Russian electricity cable issue during 2006 indicate a change into this.

The Finnish amorphousness in the energy dialogue is at least partly explained by the threat-centred narrative which has directed Finnish policies towards Russia. As it was argued, the whole Finnish narrative is constructed around the idea of Finland having to combat threats, mainly environmental, that Russia is posing to it, and facilitating co-operation between the EU and Russia is seen as one instance of this policy. Thus, it can be argued that Finland has stressed the need of the EU–Russia co-operation and the positive interdependence of the parties, best crystallised in the energy sector, mainly in order to make them to accept the idea of co-operation as a general principle. Simultaneously Finland has high hopes of facilitating environmental co-operation, but despite relative successes within the ND framework and elsewhere, the linkage from the environment to energy has remained somewhat weaker than it is in the Norwegian–Russian case. Thus, in the Finnish case energy has been used as a stool pigeon for both overall and environmental co-operation.

The Norwegian and Finnish cases represent extreme ends among the Nordics. Yet, one can conclude that the Nordic states' relative heterogeneity vis-à-vis the EU–Russia energy dialogue adds to the relative polarization in contemporary EU and EEA Europe where existing general and energy specific divisions hamper efforts of creating a pan-European energy policy (see Chapter 2). Although the northern European area has a testing ground function within the energy dialogue and includes several pilot projects (see Chapter 4), the analysis in this chapter suggests that actors in the Nordic part of the ND area confuse and question the EU–Russia energy dialogue more than they facilitate it. To this must be added the important qualifier that it is unclear to what extent small EU member states can affect the dialogue. The EU Commission too often seems to conduct the policy of the deaf in relation to them. As long as the Commission lacks full mandate in external energy issues, and adequate resources, the Nordic cases suggest that it would be advantageous to listen to the member countries and engage them more firmly into the dialogue. There is a natural division of labour in the issue: while the Commission is needed to facilitate the dialogue on the highest political level, member countries and regional actors already experienced in cooperating with Russia in energy issues remain best placed to bring concrete form to the co-operation. The Nordic countries' receptiveness in relation to engaging Russia, and their regional co-operation activities involving Russia in northern Europe can offer a lot for the energy dialogue when properly capitalized on. This necessarily means widening the energy dialogue's agenda to accommodate the Nordics' concerns.

The more general lesson for energy policy in the pan-European area and beyond is that the Nordic case in a very interesting manner illustrates the interplay of material and politico-normative considerations in energy policy. It is possible to draw a fairly direct link from divergences in material factors like resource distribution to divergences in political interests, resulting in the respective Danish, Swedish and Norwegian positions, and also incorporating their stress on the environment as coastal states with a vulnerable northern ecosystem and territorial waters. But in Finland's case the material and economic logic falls a bit short. Pure economic calculations would suggest that being dependent on energy imports, it would be rational to buy from a neighbour whose resources are abundant for the foreseeable future considering Finland's relatively low needs as a small country, although one with a cold climate and a notably high-consuming industrial sector. The decades-long reliability of deliveries from Russia compared to parts of Eastern Europe and the absence of transit countries would support this. However, the interference of politically constructed threat perceptions of energy dependence dissolves the image of materialistically and technically conducted energy policy. This remark also boils down into a lesson to the European Union in trying to handle energy policy questions with limited competencies and within a relatively polarized environment, with the Commission and many member states pursuing different agendas. In such a situation, perhaps surprisingly, focus on mere technicalities can get one only so far.

Chapter 7

The Baltic Gateway: A Corridor Leading Towards Three Different Directions?

Eiki Berg

Introduction

This chapter discusses the often suggested idea of a 'Baltic gateway' that can for its own part promote co-operation in EU–Russia energy relations. The Baltic nations are geographically at the heart of northern Europe's geopolitical lynchpin, the Baltic Sea region, giving them a chance to prove their international value in efforts of shortening the distance between the east and west in northern Europe and Europe on the whole. And northern Europe's recent history indeed demonstrates how regional co-operation can in favourable conditions bring together nations from previously different political systems that are now pursuing similar goals. But this integrative potential of the Baltic states notwithstanding, it would be wrong to assume that the Baltic nations comprise a single whole or that each nation individually behaves as a unitary actor in the myriad of international relations. For many years, Latvia has practically been the only Baltic state promoting the idea of Baltic Sea co-operation wholeheartedly while Estonia has tightened its relationship with Finland, and Lithuania has moved closer to Poland.

Different Baltic approaches also prevail in defining their role in northern Europe's energy transit business. Although the Latvians sympathize with the gateway idea of promoting openness towards all possible directions, and the Estonians together with Lithuanians point to their countries' geographical locations and historical ties within the 'north' and 'west', the three Baltic states have on the whole adopted rather mixed strategies. Liberal Estonia relies on its well-developed transportation infrastructure and ports, and the economically even less state controlled Latvia boasts a transit strategy, and exploited its oil pipeline from Russia until Russia built alternative outlets. At the same time, slightly mercantilist Lithuania benefits from its oil refinery. In this way, by providing a corridor leading towards three different directions – by means of ports, pipelines/railways, and refineries – the Baltic states have quickly differentiated and assumed mutually competitive roles.

These developments have led the Estonian capital Tallinn and the Latvian port town of Venstpils acquiring an image of a north European gateway and transit destination for eastern riches, although in the case of Ventspils, this image has in the new millennium become somewhat tainted with Russia's redirection of its transport routes. The Mazeikiu refinery in Lithuania, for its part, has become an enormously attractive enterprise given the lack of refining capacity in the region. This all hints at the capacity of the Baltic states to turn the traditionally negative aspects of their strategic importance into an advantage. Yet, as the case of Ventspils suggests, a darker side balances the otherwise positive impulses. Transit business is vulnerable, depending largely on Russia's political interests. This can reduce states' economic security, cause environmental problems and in the end produce not much common good. Given this dual picture, one should also look at the Baltic states' internal structures which also push and pull the transit corridor idea into very different directions.

The mutual competition and internal political struggles do not detract from the fact that the Baltic states function as a transit corridor and thus occupy a specific role in the EU–Russia energy link. But it would be wrong to assume that they all have crystal clear strategies and proactive stands in these questions. I would even argue that instead of looking at official positions replicating the commonplace expectations, one has to construct each state's orientation from scattered pieces of information which in each of the three cases vary a lot. This is also one of the reasons why official positions remain so obscure on these questions in the Baltics: the Balts' security concerns do not really comply with the business interests of the private sector, while environmentalists are prone to object to everything that attempts to turn the Baltic states into corridors of uncontrolled flows.

This chapter first provides a general view onto the Baltic EU–Russia energy link. Then the chapter explores in a comparative survey the roles of the Baltic states in the chain of East–West transit activities. Finally the chapter analyzes the roles of different actors involved in the transit business in the Baltic states. The focus is on the intersection of security challenges and economic interests, with the associated environmental concerns playing a more minor role in this case than in the Nordic states (see Chapter 6). Security, economic and environmental issues are shown to end up with conflicting logics. Businesspeople argue for intensive flows of transit goods with the need to increase the state's income. Policy makers point to security risks, which may result from economic dependency on foreign capital, energy carriers and service providers. Environmentalists worry about the safety standards and equipment of oil tankers and the pollution dangers in transportation activities. In this way a gateway of three states, three strategies and three set of actors competing with each other, leads towards different directions.

The EU–Russia Energy Link and the Baltic Transit Corridor

It is no secret that western Europe has suffered from chronic deficiency of energy resources already since the 1960s. In the days of global economic crises in 1973–5, caused by rocketing oil prices in the world market, EU used to import most of its oil from the Middle East, being thus extremely vulnerable and easily affected by whatever political turbulence in the area. Meanwhile, the discovered new oil reserves in the North Sea only temporarily met the demand of rapidly developing industrial countries of Europe. Even today, the EU relies on imports for almost 80 per cent of its oil supplies and about 43 per cent of its gas consumption.[1] At the same time, Russia has gained the second position in crude oil production globally after Saudi Arabia by a share of 11 per cent. What seems to be even more important is Russia's possession of almost one third of the world's natural gas reserves, making it the world's most important producer and exporter in this sector.

The oil and gas sector undoubtedly have a fundamental role in the Russian economy. Oil and gas cover ¾ of Russia's primary energy consumption and form some 15–20 per cent of the GDP.[2] Moreover, oil and gas firms are responsible for over 50 per cent of the federal budget revenues and account for over half of Russia's total exports. Russia's resources are also highlighted by the 'neoconservative alternative' which seeks to reduce US reliance on Saudi Arabia's oil supply by cooperating with Russia.[3] Being a notable exporter of raw materials will most likely help Russia to maintain its position in the world economy. But according to the former US diplomat Keith Smith, this is accompanied by Russia's attempts to grab the role of a transcontinental bridge linking east and south Asian countries with Europe while controlling the transit and export flows by using its own ports and imposing higher tariffs on foreign carriers.[4]

The persisting instability in the Middle East leads the EU to increasingly rely on Russia's energy resources. Especially this concerns the new EU members in Central and Eastern Europe, who import 80 per cent of their oil and 75 per cent of their gas from Russia, compared with 15 per cent and 20 per cent for the 'old' EU–15.[5] Given that these energy poor countries serve as transit corridors, implies that Russia may use a variety of political tools against those who are on the way of its commercial interests or foreign policy goals. Russia often stresses that state control over the nation's oil and natural gas pipelines will be a key tool for maintaining its economic and political influence beyond its borders, thus establishing itself as a great energy power in compensation for the blow it suffered to its international status when the Soviet Union collapsed.[6] In this respect, the new EU members fear the

---

1 Debra Johnson, 'EU–Russia Energy Links', in Debra Johnson and Paul Robinson (eds), *Perspectives on EU–Russia Relations* (London: Routledge, 2005), pp. 175–95.

2 Peeter Vahtra and Kari Liuhto, 'Expansion or Exodus? – Foreign Operations of Russia's Largest Corporations' (Turku Business School, Pan-European Institute, 2004), <www.tukkk.fi/pei>.

3 Joe Barnes, Amy Myers Jaffe and Edward L. Morse, 'Geopolitics of Russian Supply and US Foreign Policy', in *The Energy Dimension in Russian Global Strategy* (The James A. Baker III Institute for Public Policy of Rice University, 2004), <http://www.rice.edu/energy/publications/russianglobalstrategy.html>.

4 Keith Smith, *Russian Energy Politics in the Baltics, Poland, and Ukraine: A New Stealth Imperialism?* (Washington: The Centre for Strategic and International Studies Press, 2004).

5 'Ivan at the Pipe: Special Report on Russian Energy Firms', *The Economist*, 11–17 December 2004.

6 Isabel Gorst, 'Russian Pipeline Strategies: Business vs. Politics, in *The Energy Dimension in Russian Global Strategy* (The James A. Baker III Institute for Public Policy of Rice University, 2004), <http://www.rice.edu/energy/publications/russianglobalstrategy.html>.

special relations developing between EU and Russia and which extend far beyond commercial considerations. Out of the central and east Europeans, distrust of Russia is strong enough in the Baltics where the ex-prime ministers in 2006 warned against Russia's energy weapon which is used to impose leverage on the neighbouring countries in the west via a differentiated price policy.[7] Many other countries of the CEE region, with the notable exception of Poland, are keener to learn from the 'old' members to accommodate Russian companies approaching business with unconventional rules.[8]

According to Vahtra and Liuhto, the internationalization of Russian oil and gas majors and their entry to markets in different regions is often marked by the political ambitions of the Russian Government, pushing these 'patriots' to execute the Russian foreign policy through their foreign operations.[9] Insofar as the oil and gas industry fail to operate on the basis of the free markets principle in the exploration licenses and transportation quotas questions, the companies in the sector continue to carry out the political ambitions of the Russian Government. Similar view is presented by the political commentator Paul Goble, who links Russia's oil and gas industry directly with the state, saying that energy is a potentially powerful weapon that can be used by Russia to press its national interests and exercise political influence abroad while avoiding military confrontation with NATO.[10]

With an eye on Russia's foreign investments in the Baltic states, Vahtra and Liuhto find Lithuania on the second place among the ten new EU member states in terms of attracting Russian investments (288 million US dollars by 2004).[11] These include the acquisition by the Russian oil giant Yukos a 54 per cent stake in Mazeikiu Nafta, including a refinery and oil terminal in Butinge; in addition, Yukos also obtained a majority of stakes in the oil pipeline from the Belarusian border to the Baltic Sea, before the ownership of these facilities was questioned after the bankruptcy of Yukos (see also Chapter 3).[12] Correspondingly, Gazprom acquired a 34 per cent stake in the gas utility Lietuvos Dujos. The three biggest Russian investments in Latvia are Latrostrans (transit of oil products), Latvijas Gaze (gas supply) and Lukoil Baltija (transit of oil products and their trade).[13] These three investments (171 million US dollars) cover over 60 per cent of the Russian foreign direct investment (FDI) in Latvia. Energy companies are also behind the majority of Russian FDI in Estonia (78 million US dollars), investing mainly in gas supply and chemical production.

Surprisingly enough, this has alarmed neoconservatives in the US rather than political actors in the EU, who tend to take a pragmatic approach towards their important energy supplier.[14] Indeed, the EU today consists of 27 member states with

---

7 Delfi uudised, 'Ekspeaministrid hoiatavad energiarelva eest', 17 January 2006, <http://www.delfi.ee/archive/print.php?id=12019776>.

8 'Ivan at the Pipe: Special Report on Russian Energy Firms', *The Economist*, 11–17 December 2004.

9 Vahtra and Kari Liuhto, *Expansion or Exodus?*

10 Paul Goble, 'Julgeolek sõltub ka energeetikast', *Päevaleht*, 5 May 2005.

11 Vahtra and Kari Liuhto, *Expansion or Exodus?* p. 17.

12 Smith, *Russian Energy Politics in the Baltics, Poland, and Ukraine.*

13 Vahtra and Kari Liuhto, *Expansion or Exodus?*

14 Johnson, 'EU–Russia Energy Links'.

all having their own independent interests in energy issues. Some EU countries, especially Germany, France and Italy, view Russia as an indispensable partner necessary for increased energy supplies, which leads Keith Smith to conclude that individual European governments prefer to deal with Russia on a bilateral basis.[15] The EU's focus is on increasing energy supplies from Russia, yet this leaves relatively free hands for the member states to pursue their own agenda with Russia on a bilateral basis. As long as common EU policy in energy issues is missing, the EU Commission can only follow the game from the backbenches: for example, how the personal relationship between Russia's President Putin and Germany's ex-Chancellor Schröder develops into a deep co-operation in the business sphere, bypassing the interests of the Baltic states and Poland by projecting the Nord Stream gas pipeline from Russia to Germany (see Chapter 5). What is an even more worrisome tendency in the minds of many leaders of new EU member states is the leniency towards Russia's violation of central WTO principles such as non-discrimination and equal treatment for example in connection to Russia's differentiated tariff policy on railways. Thus many among them are asking: is it due to energy dependency and western business lobby that preferential treatment is granted in order to buy into attractive assets in Russia?

The Baltic states are tied to Russia by pipelines, rail lines and refineries. Geography alone dictates that Russia will likely remain the nearest and least costly supplier of oil and gas to the Baltics. But geographical facts may give determinists a good reason to think whether the Baltic states' position in the geopolitical crossroads should be rather taken as a challenge in terms of the prevailing power equilibrium and interstate relations. One may easily conclude that Baltic–Russian relations are in a downward curve, regardless of the fact that all three states have achieved their strategic objectives of joining the EU and NATO. As a result of these, Russia expected to witness a change in the official political discourse in the Baltic states concerning the interpretation of the Soviet period as an occupation, with claims on compensation from Russia for human and economic losses during that era. Another expectation in Russia was for a more inclusive treatment of non-citizens in Estonia and Latvia and the Russian-speaking population in general. Lithuania has had the easiest ride, although the Kaliningrad issue remains subject to potential demands from the Russian side for a corridor linking this exclave with mainland Russia.

Almost constant tensions in bilateral relations are interpreted in the Baltics as a manifestation of Russian neo-imperialism and as attempts to exert leverage in the post-Soviet space, while in Moscow the reference is to the nationalist tones expressed by some politicians in Estonia and Latvia in order to gain domestic political capital. No matter the content of specific problems ('thematic hotspots') in bilateral relations such as the question of the status of Russian-speaking minorities, unsettled border issues and trade treatment problems, transit flows and the status of the Russian Orthodox Church in Estonia, as well as property and compensational demands dating to the Soviet era, but the key point is to understand that these incidents to a large extent remain symptomatic of a wider conflict. The broader underlying conflict has to do with the incompatibility of the dominant self-conceptions and historically

15 Smith, *Russian Energy Politics in the Baltics, Poland, and Ukraine.*

based identity-narratives in Russia and the Baltic states. Both parties deny each other's narratives of the *self* and *the other* – e.g. Russia as a 'liberator' of the Balts in the Second World War vs. Russia as an 'occupier' during its course and the Soviet era – and thus problematize their role recognition by the international community.

Due to the complicated nature of the Baltic–Russian relationship, the Baltic states are often perceived as a bottleneck rather than a transit corridor. This is also something that Putin's administration knows well while forcing Russia's major energy companies to export their products through Russian ports, and thereby weakening any potential political or financial leverage over Russia by the country's neighbours. Current tariff structures mean that the cost of sending rail cargo to the Baltics is up to three times higher than the cost of sending it to St. Petersburg.[16] The construction of the Baltic Pipeline System (BPS) which carries oil to the port of Primorsk is designed specifically to bypass the Baltic states by directing most West Siberian crude oil to Russian-controlled ports.[17] Gazprom's planned Nord Stream pipeline, bypassing Baltic and Polish transit corridors, and enabling direct large-scale deliveries of Russian gas to the EU area under the Baltic Sea, is considered by the EU Commission one of the top priorities for the development of EU–Russia energy link.[18]

The idea to build a gas pipeline from Russia to Germany along the bottom of the Baltic Sea was set into motion in order to balance sovereignty risks and avoid transit costs. Being the second-longest undersea gas pipeline in the world (1189km) it will run from Vyborg and end in Greifswald. Although many in the Baltic States wished to draw historical parallels with the Molotov–Ribbentrop Pact which prior to the Second World War divided northern parts of eastern Europe into German and Soviet spheres of influence,[19] hence terming the project a Putin–Schröder Pact, these must be seen as alarmist voices in a wrong time and wrong place. Preparations for the project started already in 1997, and received a high level of support and recognition from EU institutions in 2000. In contrast to the alarmist interpretations it can be argued that as long as there are consultations with all the coastal countries of the Baltic Sea – and provided that there will not be any fluctuations in gas supply, and that environmental issues will receive the needed attention – then energy security will not be weakened for any directly or indirectly involved parties.

Transit Business in the Baltics

Although the Baltic states lie geographically at the heart of the Baltic Sea region, they rely on different partners and strategies to prove their international value at the face of the international community. Only Latvia as a 'truly Baltic state' has positioned itself in the very centre of the region and opened its borders to all different directions.

16 B. Nimmo, 'Russia's Ports a Challenge, Not a Threat', *The Baltic Times*, 18 February 2005.

17 Smith, *Russian Energy Politics in the Baltics, Poland, and Ukraine.*

18 Vahtra and Kari Liuhto, *Expansion or Exodus?*

19 Delfi uudised, 'Poola minister võrdleb gaasijuhet MRP-ga', 3 May 2006, <http://www.delfi.ee/archive/article.php?id=12801526>.

The contrast to Estonia is striking. Its EU accession became a desperate attempt of '*drang nach Norden*' in a series of branding exercises of portraying the country as a 'post-communist Nordic state', with a Scandinavian style cross-flag and in some accounts, even 'Estland' as its proposed new state name, thus further linking it to Finland and the Nordic group.[20] Lithuania's close association with Poland sparked the idea of its route to 'Europe' going through Warsaw.[21] This neglected the obvious fact that Lithuania, like Poland, is historically *both* a Baltic and a Central European state, and that her interests require her to keep both avenues open.

At the same time it would be wrong to argue that the gateway idea was something that the Latvians only sympathized with. There were also others who promoted openness in all possible directions without nailing up the eastern window. For instance, the 1990s foreign policy statements in Estonia pointed to its geographical location and historical ties that make the country a 'bridge' or a 'land of contacts'.[22] Some commentators even identified Estonia as a southern part of Finland due to the fact that the 78km distance between capitals Helsinki and Tallinn had almost disappeared because of intensive traffic and communication flows.[23] Indeed, by the end of the 1990s, Baltic cities became important gateways and transit destinations for eastern riches, thus supporting the idea of giving some positive input into the mediation of business between the east and west.

Yet, the lack of resources conditions the Baltic states' economic choices. On the one hand, they can rely on their geographical location and human resource potential, and on the other, they can exploit their relatively cheap labour and manufacture low-cost products for the world market. As for Estonia, there are also analysts who foresee its role in the world economy as a service provider: to be involved and useful in the east–west contacts as much as possible.[24] Latvia, by contrast, more clearly positions itself as a country prioritising east–west links. Latvia's National Transport Development Programme (1996–2010) portrays Latvia as a transport corridor in-between resource-rich Russia and Central Asia, and resource-poor western Europe. In fact, Estonia's and Latvia's geographical position and the structure of flows in transit business makes them quite similar at the outset until the Russian oil companies' bypassing of Latvia's Ventspils port. Estonia's comparative advantage is its efficiently functioning seaport, and Latvia benefits from its well-operating railway system.[25]

20 Eiki Berg, 'Some Unintended Consequences of Geopolitical Reasoning in Post-Soviet Estonia: Texts and Policy Streams, Maps, and Cartoons', *Geopolitics*, 8/1 (2003): 101–20.

21 Thomas Lane, 'Lithuania: Stepping Westward', in David Smith, Artis Pabriks, Aldis Purs and Thomas Lane (eds), *The Baltic states: Estonia, Latvia and Lithuania* (London: Routledge, 2002).

22 Toomas Hendrik Ilves, 'Vabariigi Valitsuse nimel peetud ettekanne Riigikogus', 12 February 1998.

23 *Eesti tulevikustsenaariumid: üleriigiline territoriaalmajanduslik planeering 'Eesti 2010'* (Tallinn: Eesti Vabariigi Keskkonnaministeerium/Eesti Tuleviku-uuringute Instituut, 1997).

24 Akadeemilise Balti ja Vene Uuringute Keskus, 2004.

25 *Eesti transiidikoridori konkurentsivõime ja selle tõstmise võimalused* (Tallinn: Strateegiliste Algatuste Keskus, 2003/2004).

Finnish experts have come to a similar conclusion according to which transit traffic to and from Russia and the CIS countries is one of the most important and most profitable sources of income for the Baltic states, especially to Latvia and Estonia. The geographical location of the Baltic ports makes them favourable to the various centres of Russia. Being largely ice-free and with deep docks, these ports are well connected with railway networks, terminals and warehouses. Ports have also skilled operators and personnel with good command of the Russian language. But, whether the Baltic states maintain the economic gateway position or develop this even further depends mainly on Russia's sectoral policies, EU's energy dependency and strategic planning in the transit countries themselves. This Finnish report even predicts that the significant Russian investments into the infrastructure and terminal operations in the Baltic ports may contribute to a further use of these ports.[26]

A clear majority of transport by rail is transit of oil products via Baltic ports. As said, the oil pipeline to Ventspils remains unused due to Russia's policy of favouring its own ports. Transit volumes by rail are growing especially via Estonian ports.[27] During 1995–9, on average 56 per cent of Estonia's total transports were Russian transit, and by 1999 oil had assumed 70 per cent of these, making up about 10 per cent of the GDP. Underlying the growth of oil transit trade were Estonia's improvements on railway transports and logistics, expansion of the Muuga oil terminal near Tallinn and the good co-operation between the Estonian Pakterminal oil shipment company and Russia's Lukoil.[28] According to Juhani Laurila, these factors helped Estonia to grab 9 per cent of the value of Russia's westbound transit by 1999.[29] However, Russia's port-building projects around St. Petersburg increase Russia's own transport capacity and by 2010, may reduce the Baltic transit by 75 per cent.[30] Estonia's transit potential may be under threat, although the projected increase in Russia's export volumes into the EU, combined with the diversification of the freight structure, may work to offset these threats.

At the first glance, Latvia has more to offer for east–west transit flows when compared to the other Baltic states. Ventspils is the only port in the Baltic States with a direct oil pipeline connection from Russia. Ventspils and the other Latvian ports, Liepaja and Riga, managed to grab almost 30 per cent of the total Baltic Sea cargo in 1999.[31] Furthermore, according to the estimates, Latvia's share of the value of Russia's westbound transit was 26 per cent, whilst figures for the importance of the

26 *Transport Connections Between the EU and Russia: Current Status and Outlook for the Future* (2005) (Helsinki: Ministry of Transport and Communication Finland), <www.mintc.fi/oliver/upl255-Julkaisuja%2010_2005.pdf>, accessed 31 July 2006.

27 L. Ojala, T. Naula and T. Hoffmann, 'Trade and Transport Facilitation Audit of the Baltic states: On a Fast Track to Economic Development' (The World Bank, 2005).

28 Pami Aalto, *European Union and the Making of a Wider Northern Europe* (London: Routledge, 2006), p. 63.

29 Juhani Laurila, 'Transit Transport Between Russia and European Union in Light of Geopolitics and Economics', *Emerging Markets Finance and Trade*, 39/5 (2003): 27–57.

30 David Smith, 'Estonia: Independence and European Integration', in David Smith, Artis Pabriks, Aldis Purs and Thomas Lane (eds), *The Baltic states: Estonia, Latvia and Lithuania* (London: Routledge, 2002).

31 Aalto, *European Union and the Making of a Wider Northern Europe.*

transit trade for the Latvian economy suggest that in 1998, transit accounted for 8–10 per cent of Latvia's GDP.[32] But things turned different in 2002, when the owners of the port of Ventspils rejected an acquisition offer from Transneft and Lukoil. Almost immediately, Transneft announced that no Russian crude oil would be carried by the pipeline to Ventspils until a sale was negotiated that would give a majority of shares to a Russian company.[33] At the background were Latvia's problems in relation to Russia, eventually resulting in a total cut of supplies to Ventspils.[34]

Lithuania has much to gain from transit trade since its good infrastructure and transport potential supports the intersection of major east–west and north–south transport corridors.[35] Two out of the nine EU-designed trans-European transport corridors pass through Lithuanian territory. The ice-free Lithuanian port of Klaipėda offers one of the cheapest routes from the Baltics to Belarus, Ukraine and southwest Russia. But Klaipėda's comparative advantage in relation to Estonian and Latvian ports is its lesser reliance on Russian markets: about 80 per cent of the port's transit goes to and from Belarus and Ukraine. Hence, out of Russia's westbound transit trade, in 1999 Lithuania accounted for only 6 per cent. The ensuing transit income made up an estimated 4–6 per cent of Lithuania's GDP.[36] Lithuania's bridge function is largely supported by its oil infrastructure. It connotes the energy complex consisting of the Mazeikiu oil refinery, where crude oil can be supplied through the Būtingė oil terminal, thus reducing further the dependence on Russian supplies (see Figure 7.1).[37]

Despite of the transit potential of the Baltic states and its full exploitation in the context of EU–Russia energy links, there are at least three types of concerns for the Balts. The first one has to do with the ownership of infrastructure. The port of Ventspils used to function as a profit-maker for the city's Mayor and his business associates with close ties with Lukoil. There are good grounds to believe that any new deal opening the oil taps will have a visible Russian involvement. And the Lithuanians, as phrased by *The Economist*, wanted to 'keep Ivan away from the pipe', but Ivan was already there, since Mazeikiu Nafta relied entirely on Russian oil, most of it from Lukoil, which itself was eyeing to buy the refinery.[38] The US company Williams which invested first into the Mazeikiu oil refinery in 1998, experienced setbacks in Russia's oil supply, and sold its shares to the Russian oil company Yukos. The shares owned by Yukos and the Lithuanian Government shares were eventually bought in December 2006 by PKN Orlen, giving the Polish company an 84.4 per cent ownership in Mazeikiu Nafta.[39]

---

32 Laurila, 'Transit Transport Between Russia and European Union'.

33 Smith, 'Estonia'.

34 Aalto, *European Union and the Making of a Wider Northern Europe*.

35 Ibid.

36 Lane, 'Lithuania'; Laurila, 'Transit Transport Between Russia and European Union'.

37 Aalto, *European Union and the Making of a Wider Northern Europe*.

38 'Ivan at the Pipe'.

39 See Gary Peach, '"Deal of the Century" Finalized – PKN Orlen Buys Mazeikiu Nafta for $2.3 Billion', *The Baltic Times*, 20 December 2006.

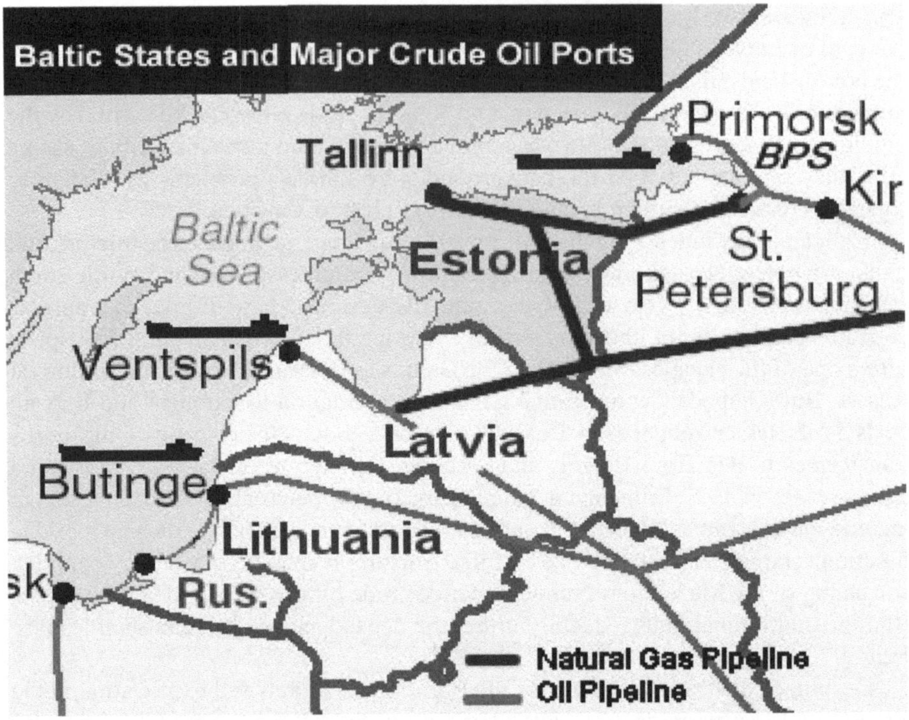

Figure 7.1 The Baltic transit corridor

Source: Energy Information Administration (EIA).

In Estonia, where transit trade takes place through railways, the partly US-owned Estonian Railways is losing its monopoly status and therefore faces the prospect of giving up its control over railway infrastructure due to the EU's competition policy. The most interested parties looking for accessing these facilities are again in Russia. However, the Estonian Government 'renationalized' the 66 per cent of the shares in Estonian Railways shares that formerly belonged to the US based company Baltic Rail Services, upon the company deeming the tariffs set by the Estonian Government for Russian transit yielding too little revenue. But on the whole, it is too early to say whether these ownership structure developments in Estonia, Latvia and Lithuania indicate a strategic shift in the bigger picture of the EU–Russia energy link and the role of the Baltic states within it.

The second Baltic concern relates to how both well-established and new investors of Russian origin represent a clear interest of the Russian state to exert political leverage over the Baltic states. The struggling over the control of the Ventspils pipeline, Mazeikiu Nafta and even Estonian Railways give an indication of what sort of issues are implicated in Russia's assets buying strategy and economic involvement in the former Soviet Union area. Even though Russian actors may not always succeed in their aims, it is worth mentioning that Russia has used energy policy weapons against the Baltic states in the past, and it is claimed that it looks to use these now, whether openly through propaganda and diplomacy, or covertly through espionage.[40]

The third Baltic concern is that despite their potential role as a gateway in the EU–Russia energy link, each Baltic state in its own way seems to have lost faith with EU solidarity when geopolitical issues such as safety of energy supplies are at stake. Although having a voice within the EU, the Baltic states have not been able to convince the other member states to stop Russia boosting the capacity of its own ports for mere political reasons at the expense of sound economics. Nor is there entirely unambiguous information on what conclusions the EU drew after the Russian–Ukrainian gas dispute that emerged in the beginning of 2006: for example, to bypass Ukraine and continue reliance on Russia's energy sources, or to diversify energy supplies?

Security Concepts and Strategic Visions[41]

The Baltic states' position within the larger EU–Russia energy game and the three related Baltic concerns that were outlined above suggest a strong presence of security political considerations in their transit policy. These must be first examined for each country in more detail before the transit trade actor structure in the Baltic states can be fully spelled out.

As for Estonia, regardless of the share of transit in the country's GDP, most opinion leaders in the political sphere express distrust in the infiltration of Russian capital. Economic interests derive from projected profits, which for their part motivate people to do whatever things, is a high official from the Estonian Ministry of Foreign Affairs convinced: 'in my view, Estonia is vulnerable to outside economic pressure and that naturally affects national security'. The Latvians are more relaxed, as the Chief Executive of the Ventspils Free Port puts it: 'the way I look at Russia is that we are partners and hopefully business-to-business relations remain based on rational and correct principles. Whoever is the owner of the company operating the port is a question I couldn't care less.' Also the Director of Lithuanian Railways sees nothing bad or strange in Russia's business activities because 'that is exactly what they are for – to earn profit. The bonus they have in this region is that they can play one Baltic state against the other and thus have more influence on each Baltic state individually'.

The document 'Security Policy Principles' adopted by the Estonian Government in 2004 became the one where transit related risks were for the first time mentioned in terms of economic security, with the assertion that a major threat factor is the great dependence of Estonia's gas and electrical systems upon foreign monopolistic energy systems and suppliers. Accordingly, it is in Estonia's interest to promote the diversification of external economic relations. In the context of the development of east–west transit trade, it is possible to succeed despite increased competition

40 'Ivan at the Pipe'.

41 This section makes use of interviews conducted during 2004 among Baltic transit trade actors, including politicians, officials, business representatives and environmentalists.

and political influence being exerted upon economic processes.[42] Transit is taken as an important economic sphere, but due to the associated national security issues, it is not considered without reservations. When speaking of the 'eastern threat' in the economic sense, the policy planners in the Estonian Ministry of Foreign Affairs would prefer to know exactly whom they are dealing with: 'if the market share of local companies (the ones that are partly state owned) is threatened and if foreign companies acquire some 30–40 per cent of the market share, then the state should definitely intervene'. Although in official rhetoric not very often heard, the Estonian Ministry of Foreign Affairs continues to have a fairly negative stand in relation to the country's dependencies and its perceived vulnerability in the neighbourhood of its former patron Russia.

Transit is not officially promoted by the foreign policy elite but rather left to struggle for better conditions by itself. This is what service providers in the Estonian Railways think: 'if we have a chicken that lay golden eggs then we should not kill the chicken nor let somebody to steal the eggs.' Estonia's Ministry of Economic Affairs and Communications is positively tuned, meaning that transit is an important economic sector to be supported also politically. It has developed a vision for the period 2006–13, according to which there is a need for a transport system which satisfies the movement of people and goods, and at the same time is efficient, safe and environmentally friendly. This development plan refers to Estonia's transportation infrastructure – service providers and logistical units competing successfully with Finland. Weaknesses are in the structure of goods (mainly oil and oil derivates) and their unidirectional transit flows (east–west route dominating). But the plan also sees the risks of being dependent on Russia's economic policy and notes the environmental concerns ensuing from the massive oil transfers.[43] Contrary to Latvia and Lithuania, one of the high officials claim, 'transit activities in Estonia do not create added value since the networks of techno-parks and logistic centres are poorly developed.'

The recent development scheme in transportation for 2006–2013 has a strong focus on Estonia's transit sector. Indeed, this document enlists activities that the state can do in order to keep momentum in the development of the sector, as well as assess the impact of the strategic environment on the transit business. But it does not contain a strategic view. The director of the Union of Port Operators of Estonia is convinced that 'we have not defined our economic interests while doing business with Russia nor is it conceptually crystal-clear whether we need transit at all. Our Foreign Ministry's view is that less transit flows increase Estonia's security'. Apparently, transit partnership with resources rich Russia develops only on the level of businesses. This view is expressed also by other Estonian businessmen active in transit sector: 'we do not have a clear-cut vision for the transit sector. Nobody has ever used a legal channel to influence a process because according to the Foreign Ministry it does not sound good'. The Estonian think tank, Centre for Strategic Initiatives has reported that the competitive entities for enlarged markets are no

42 'National Security Concept of the Republic of Estonia 2004', <http://www.vm.ee>.

43 Majandus- ja Kommunikatsiooniministeerium, 'Transpordi arengukava, 2006–2013', 2005, <http://www.mkm.ee>.

longer firms but states (or transit corridors crossing states).[44] State involvement in transit business is a growing trend in Latvia, Lithuania, Finland, and especially Russia, but not in Estonia.

This suggests that Estonia needs a comprehensive transit strategy, which is currently missing. Nobody has actually drafted a strategic document on the governmental level, tackling transit flows, competing transit corridors, Estonia's comparative advantages, etc.[45] Estonian businessmen commonly think that the 'honeymoon is over', and that 'we need to think analytically where we are situated in the world's transit business and what would be the state's role in leading the process, or we simply have two options: either we do not deal with these transit issues and lose the whole business or we have a position within it while keeping certain rules of the game in mind'. To the disappointment of many, the Transit Committee in Estonia is an advisory body, a talk-shop meeting regularly and discussing issues that are mainly technical and rarely policy affected. Interested parties seem to prepare the agenda without much expectations to end up with a working document or an analytical piece.

Latvia, on the contrary, has more clearly identified itself as a country contributing to the east–west business. The country's National Development Program (1996–2010) considers Latvia as an east–west transit corridor relying on European, Russian, Belarusian and Baltic economic interrelationships and transportation networks. Among the Balts, the Latvians are the ones who prioritize Baltic Sea regional co-operation and have promoted the idea of an 'Amber Gateway' with its political and economic promises.[46] This kind of reasoning explores the market potential of an area with roughly 80 million people, being one of the most vital economic regions of Europe. The economic development of the country is also related to the ability to maintain and increase the existing flows of transit cargo. According to the 2002 National Security Concept of Latvia, the main tasks in the development of transport services are provision of competitiveness in services, development of a unified transport infrastructure for inland and international transport, development of Latvia's port infrastructure, development of oil and oil product pipelines, modernisation of the road network, the east–west railway corridor and related infrastructure. Latvia is committed to continue active participation in the expansion of the trans-European transport network.[47] However, Latvia's practical record is worse than Estonia's despite the considerable efforts for the sector's development from the part of the state.

In the Lithuanian case, the transit sector is framed in security terms. Lithuania's 2002 National Security Strategy depicts the overwhelming dependence on the strategic resources and energy supplies on one country (read: 'Russia') as a potential

44 *Eesti transiidikoridori konkurentsivõime ja selle tõstmise võimalused* (Tallinn: Strateegiliste Algatuste Keskus, 2003/2004).

45 Ibid.

46 Artis Pabriks and Aldis Purs, 'Latvia: the Challenges of Change', in Artis Pabriks, Aldis Purs, Thomas Lane and David Smith (eds), *The Baltic states: Estonia, Latvia and Lithuania* (London: Routledge, 2002).

47 'The National Security Concept of Latvia', 2002, <http://www.mfa.gov.lv>.

danger. In order to ensure the national security of Lithuania in the economic issues connoting strategic importance, it is prescribed which strategic objects must belong to the state by ownership right, and to which objects private national or foreign capital will be allowed access, provided that the controlling decision making power is retained by the state.[48] The Director of Development of Lithuanian Railways has a comparison to draw from: 'We already have the precedent in Estonia where the owner of the goods has the opportunity to say to a private company that the only thing they want is the track and that they will bring in their own operators. My feeling is that we have to co-operate more in the Baltics and promote the idea of Rail Baltica being independent of outside influences – meaning also that goods can be taken to Europe via rail and not via Russian ports'. Indeed, Rail Baltica may become one day a high-speed, European-gauge rail link from Warsaw to Tallinn. Conceived in 2003, the initiative was included as a priority in the European Commission's 'TEN-T' trans-European transport plan due for completion by 2016.[49]

During the most troubled years of Baltic–Russian relations (1994–8) the transit figures grew drastically because Russia needed to export and the Baltic states had the capabilities to satisfy the demand. However, for a long time in the Russian side, transit was considered as close to treason, saying that Russian businessmen betray their country's interest by moving goods through the Baltics instead of using their own national ports. In 1998, Russia began to redirect the transit flows more to its own Baltic Sea ports with the help of a differentiated tariff policy on railways. Today, Russia seeks to administer infrastructure in the transit countries. In the view of the Director of Development of Estonian Railways, 'European legislation condemns monopolies in favour of liberalizing markets – this means opening the door for Russian capital to acquire access to transit services via railways, thus adding a missing link between the resource extracting unit and product exporting unit'. This kind of vertical integration already took place in Russia's gas sector.

In sum, in the Baltic eyes security threats seem real, but in the present context of the EU–Russia energy links, the Baltic states remain unable to diversify the structure of transit flows, currently dominated by oil products, or, to redirect the flows from west to east. Further limitations include the fact that Russia's economic model does not favour local entrepreneurs, something that liberal Estonia and Latvia mostly have on offer, and that they have to comply with European competition policy. However, the Marketing Director of Tallinn Port may have a point when saying that 'Russian capital in the infrastructure or operating activities is the best security guarantee for purely pragmatic reasons: nobody wants to lose the money in case the negative scenario is realized'. Also the fact that the Baltic states' ports are ice-free and navigable, speaks for them. In the end, the further east one goes, the more difficult the conditions there are; thus it remains that oil keeps flowing downhill.

But one should not forget the domestic debate, where people operate with pre-existing attitudes by saying that trains make noise, the cargo stinks, or that Russian capital is unwelcome. Almost whatever reference to Russia's economic activities

48 'Lithuanian National Security Strategy 2002', <http://www.kam.lt>.

49 B. Nimmo, 'Rail Baltica, Via Baltica Projects Still Far Beyond the Horizon', *The Baltic Times*, 23 March 2005.

in the Baltic states is bound to increase the political capital of any orator, and make the electorate to perceive the transit businessmen as selfish and acting in the alleged neo-imperial interests of Russia. At the same time, the Director of Development of Estonian Railways has strong reasons to believe that 'security risks are maximized at once, when the involved people are afraid of defending their own economic interests and leave their competitors free hands to occupy the market'. A 'checks and balances' approach is what he would like to see as the Baltic way of benefiting from the EU–Russia energy links.

Environmental Concerns

According to estimates, transportation is the reason for 45 per cent of the oil found in the sea, while offshore production is source only to a 2 per cent of oil related pollution.[50] The fact that about 2000 large ships and tankers sail on the Baltic Sea every day, makes environmentalists consider the risks of major oil spills and accidents of highly explosive and toxic cargo outweighing the economic benefits. An activist from the Estonian Green Movement claims 'that transit business benefits only a handful of people, and it is neither economically nor environmentally sustainable. There is no in-depth assessment of the actual economic benefits of increasing transit flows'. Not surprisingly, the environmentalists' biggest concerns relate to leakages from tankers and to possible other accidents. The Baltic Sea is a very sensitive eco-system due to low amount of fresh waters circulating there through the Danish straits and three times lower salinity compared with the Atlantic Ocean. Despite of the EU-induced ban of single hull tankers in 2003, these vessels can operate in the international waters of the Gulf of Finland until 2015, as Russia has no obligation to follow.[51]

The first major oil spill in Estonian waters took place in September 2000 when the single hull tanker Alambra, sailing under a Maltese flag, spilled 240 tons of oil into the sea. It took four years for the Estonian authorities to publicly regret for failing to detect the oil spill promptly and remove the waste without external help.[52] In January 2006, a new spill was detected off the coast of northwest Estonia from an unidentified oil tanker. Estonian officials were heavily criticized for their slowness in reacting, both in terms of clean-up and investigation. Then in March 2006, another oil spill occurred when a Dominica-flagged cargo ship collided with another vessel and sank. The ship was carrying 102 tons of heavy fuel, 35 tons of light fuel oil and

50 A. Kostianoy, S. Lebedev, K. Litovchenko, S. Stanichny and O. Pichuzhkina, 'Oil Spill Monitoring in the South-eastern Baltic Sea', *Environmental Research, Engineering and Management*, 33/3 (2005): 73–9.

51 Elina Pelto, 'Environmental Risk of the Increasing Oil Transportation in the Gulf of Finland', in Kari Liuhto (ed.), *Growing Russian Oil Shipments in the Baltic Sea: Strategic Decision or Environmental Risk?* (Lappeenranta: Lappeenranta University of Technology Publications, 2003), p. 202.

52 'Nation Unprepared for Potential Oil Spill', *The Baltic Times*, 1 December 2004, <http://www.baltictimes.com/print_article/11506>.

600 litres of lubricant oil in its tank.[53] According to estimates these two oil spills caused the death of more than 10,000 birds.

In Estonia, the three main actors who are actively engaged in environmental protection are the Estonian Green Movement, Estonian Fund for Nature, and the Estonian Council of Environmental NGOs. Among them, the Estonian Green Movement has become one of the most influential environmental groups in the country. Its mission is to respond to the regional environmental problems brought about by the political and social changes.[54] Estonian Fund for Nature operates in close co-operation with the global network of World Wildlife Fund. It aims at preserving the rich natural diversity through active co-operation with individuals, business enterprises, organizations and state institutions. The other priorities include the promotion of sustainable resources use as well as raising public awareness of environmental issues.[55] Estonian Council of Environmental NGOs (ECEN) is an informal co-operation network 11 full members across the NGO sector. According to commonly agreed principles, member groups work jointly on the national level and have an interest in (primarily national-level) matters of environmental policy, e.g. watchdoging and lobbying the government.[56] In a press release in January 2005, ECEN criticised the Estonian Government for not paying enough attention to the preparation of crisis management forces for accidents with oil tankers causing heavy pollution of the sea and shores.[57]

Because of institutional fragmentation and weak position of the civil society, these messages rarely reach the authorities. However, all Estonian environmental organizations strive at influencing legislation and expect to influence policy-making in relevant ministries. Much attention has been paid to promoting sustainable development and public awareness of environmental concerns. Unfortunately, transit issues come only in the end of the organizations' list of priorities. This is explained by the fact that there is a constant shortage of human and material resources to tackle these problems. Also the organizations' chance to influence the direction and scope of transit flows remains minimal. As the sad acknowledgment by an activist in the Estonian Green Movement goes: 'environmentalists' hands are too short to have any significant impact on shaping the policies in the transit sector'.

Yet, an initiative has been made together with the Coalition Clean Baltic (CCB), which is the network of 27 environmental citizens' organisations, operating on grass-root level in 9 countries bordering the Baltic Sea and representing more than 500 000 members of its member organisations. A CCB Resolution which was submitted to the Baltic Sea prime ministers meeting in Estonia in June 2004 urged to give proper mandate to the relevant ministers and administrations in order to develop and implement appropriate measures for prevention and mitigation of negative

53 Kairi Kurm, 'Estonia's Oil Transit Threatens Baltic Sea'. *The Baltic Times*, 3 May 2006, <http://www.baltictimes.com/print_article/15299>.

54 Eesti Roheline Liikumine, <http://www.roheline.ee>.

55 Eestimaa Looduse Fond, <http://www.elfond.ee>.

56 Estonian Council of Environmental NGOs, <http://www.eko.org.ee>.

57 '2004 KARUTEENE ja ROHELISE TEO nominendid', ECEN Press releases 28 January 2005, <http://www.eko.org.ee/?node=3>.

environmental impacts of maritime transport of hazardous cargos. The threat perception of this coalition highlighted how maritime transport of oil products and other hazardous cargos through the Baltic Sea has doubled in the last decade, thus increasing the risks of severe accidents with oil tankers.[58]

While Estonian environmentalists oppose strongly the current trends in the transit sector, the Latvians and Lithuanians seem to enjoy a partnership with the authorities. In their party programme from 2002, Latvian Green Party wanted to make use of the favourable geographic disposition of Latvia in manufacturing and transit sectors.[59] With this in mind, the party largely conforms to the dominant political discourse of Amber Gateway and pays little attention to the negative side effects this corridor may cause to the environment. Also, one of their actions, which gained much publicity, took place not in Latvia but in Lithuania when their activists occupied the Lithuanian Oil Terminal at Butinge, and attached themselves to the buoy with handcuffs, thus trying to prevent the loading of the first oil tanker of the day at this terminal.[60] One should not forget that this Latvian–Lithuanian border region has been long disputed because of the undersea oil reserves.

Similar events took place in Lithuania where the Green Movement, instead of protesting against Mazeikiu Nafta, mobilized to target the Russian oil company Lukoil's preparations to open the D–6 oil deposit, situated 22 kilometres from the coast in the Kaliningrad region. The Green Movement is raising serious concerns about the planned oil extraction where the Russian Government is alleged to violate international agreements.[61] The raised concerns sound valid but at the same time hypocritical. Lithuania has its own role to play in the EU–Russia energy link, while the actions of its environmentalists have proved fairly toothless in preventing the creation of corridors of uncontrolled flows. As it appears, in Lithuania, environmental issues are tackled selectively and even national pride in the disputed aquatorium may turn out determining the political actions of the 'greens'.

Conclusion

This chapter proceeded from an examination of the EU–Russia energy link where the EU relies increasingly on Russia's energy supplies, and where Russia correspondingly seeks to increase the security of its energy markets. Here, the otherwise often so anonymous EU transforms into a set of independent interests expressed by member states in energy matters, while the activities of Russian energy companies are often perceived to translate to the state's foreign policy conduct. The new resource-poor member states of the EU serve the role of transit corridors where Russian actors use

58 Coalition Clean Baltic (CCB) 'Resolution', 2004, <http://www.ccb.se/doc/resolutionbaltic2004-final.doc>.

59 Latvian Green Party, <http://www.zp.lv>.

60 Ibid.; 'Latvian Greens Occupy the Buoy of the Butinge Oil Terminal', press release, <http://www.zp.lv>.

61 'Environment Ministers and the European Commission Must Stop Helsinki Convention Violations in the Baltic Sea', press release, 20 June 2003, <http://www.zalieji.lt>.

a mixed variety of political tools against those who conflict with their commercial interests or foreign policy goals.

The Baltic states have a clear transit potential while at the same time being the weakest chain in the energy link. The metaphor of a corridor leading towards three different directions reveals the mixed infrastructural potential and comparative advantages (ports, pipelines/railways, refineries), political strategies (laissez fair, moderate, mercantilist) and different sets of domestic actors that are found in these states. In the final analysis, Estonia and Latvia are quite similar along the structure of flows in the transit business. At the same time Estonia's comparative advantage is its efficiently functioning seaport, while Latvia benefits from its well-operating railway system. Lithuania gains a bit from everything but the decisive factor is its oil refinery. Because of Russia's interests to exercise an influence on political processes within the region and the EU's interests for a secure energy supply, the Russian party first tries to impose full control over the energy flows through the Baltics and then influence policy-making into a desired direction. What has happened with the Ventspils pipeline and the Mazeikiu refinery may come true also in the case of Estonian Railways, given that a laissez fair approach has minimized the state's intentions to draw a strategic vision weighing challenges and opportunities in relative terms.

The Baltic corridor also points to different directions when we approach the policy-making level. Service providers have strong economic arguments for the increase of transit flows. They dream of the prioritization of the field within the national economies given the size of GDP it annually makes up. In Estonia, many businessmen want to see a comprehensive transit strategy, while the Latvians have such a document but do not necessarily do any better. Policy-makers in all three Baltic states emphasize security risks, which may result from the economic dependency on foreign capital, energy carriers and service providers. Yet, the Latvians are somewhat surprisingly the least concerned, the Estonians do not speak about the 'Russian threat' openly, while the Lithuanians define carefully their strategic assets and decide on ownership questions, and on who gets what, how and why. Environmentalists have only limited capabilities to act in the Baltic states, and if they do act, then most likely not in the field of transit business. Especially in the Latvian and Lithuanian cases it is interesting to see how the environmental concerns mix up with national interests and economic power. In the end, oil keeps falling downhill through the Baltic corridor despite of the road signs which point to different directions.

Chapter 8

Actors in Russia's Energy Policy towards the EU

Stanislav L. Tkachenko

Introduction

This chapter examines different interest groups within the Russian Federation that play significant roles in the shaping of national energy policy towards the European Union and the external markets of Russia's energy products in general. Actors under consideration include the president and the presidential administration, ministries of the Russian Federal Government, regional administrations in the North-Western Federal District of Russia, legislative assemblies of Russian regions, leading energy companies, and other business and financial actors.

The viewpoint taken in this chapter presents the mechanisms of federal energy policy formation as a result of competition and even intriguing between different ministries, interregional relationships within Russia, and competition between companies. Special emphasis is put on analysing business and other strategic interests of Russian energy companies in conditions of high oil and gas prices, and on the prospects of attracting investments into the development of new oil and gas provinces in Russia's northwestern regions.

I will first argue that out of the various actors in Russia's energy policy, the presidential administration has played the leading role in the shaping of national energy policy with regard to both public and private segments of the energy market during Vladimir Putin's presidency. Second, I will suggest that the growing role of the Federal Government in the energy market in Putin's Russia is related to the preparations for the parliamentary and presidential elections in 2007–8. Third, energy, especially natural gas and oil, now represents the most effective and 'civilized' power political instrument for the Russian Government in the international arena. For many years to come, it will stay as an important tool in Russia's bargaining with its European neighbours in the creation of a new security political and economic architecture for Eurasia.[1] Moscow will use this instrument in a very pragmatic way, converting Russia's stable supply of energy to European markets into financial resources and growing economic integration with its most important neighbouring countries, the EU members. At the same time, relations with former Soviet republics will rather be directed towards concessions in the control of pipelines, investments

1 Irina Isakova, *Russian Governance in the Twenty-First Century: Geo-strategy, Geopolitics and Governance* (London: Frank Cass, 2005), p. 28.

into energy infrastructure and, as final result, towards total economic control by Russian public and private companies of major energy assets in the CIS area.

In order to make the argument I make use of interviews with anonymous representatives of the business community of St. Petersburg and the Leningrad region, regional legislators, and officials of Russia's northwestern regions. The analysis also derives from media sources including federal and regional newspapers and websites. Further sources include official publications of regional authorities as well as existing literature on the energy policy of the Russian Federation and the role of energy in the Russia–EU dialogue, especially after the fifth enlargement of the EU in May 2004.

Energy Sector in Russia's Economy and Politics

It is difficult to overestimate the role of the energy sector in Russia's economic system. According to estimates by John Grace, about 47 trillion cubic meters of natural gas, making about one quarter of the world's natural gas reserves, as well as 100 billion barrels of oil (13 per cent of world reserves) is located within Russia's borders.[2] The energy sector brings a major part of currency earnings to Russia's budget and is the key engine driving the country's fast economic growth since the turn of the millennium. Public and private energy companies play a very important role in the economic transformation of many Russian regions thanks to oil and/or gas fields located on their territories and their control of transportation means via their territories to export terminals. The Russian Federation inherited from the Soviet Union 46,000 kilometres of oil pipelines, 15,000 kilometres of pipelines intended for oil products as well as the world's largest grid of gas pipelines comprising altogether 152,000 kilometres. All of these pipelines are in state ownership or at least under total functional control of public institutions (Figure 8.1).[3]

The huge inflow of hard currency into the state budget due to energy exports gives Moscow the opportunity to pay back all Soviet and post-Soviet foreign debts, to enlarge the country's stabilization fund and the currency reserves of the Central Bank of Russia, to maintain a non-deficit budget for many years, and guarantee the stability of the rouble's exchange rate. In one way or another, it can be said that the whole Russian population, its business community, and federal and regional authorities benefit from the national energy sector.

During the first 15 years of Russian history since the disintegration of the Soviet Union, the relationship between governmental institutions and other actors within the energy sector evolved considerably. The starting point was in 1991, when the domination of the government in the sector was total. The state was responsible for the elaboration of strategy and the tactics of the energy sector's development; it was the only investor, and it was responsible for mobilizing material and human resources.

---

2 John Grace, *Russian Oil Supplies: Performance and Prospects* (Oxford: OUP for the Oxford Institute for Energy, 2005), p. 213.

3 Edward C. Chow, 'Rossijskie truboprovody: Nazad v buduschee?' *Pro et Contra*, 8/3 (2004): 164.

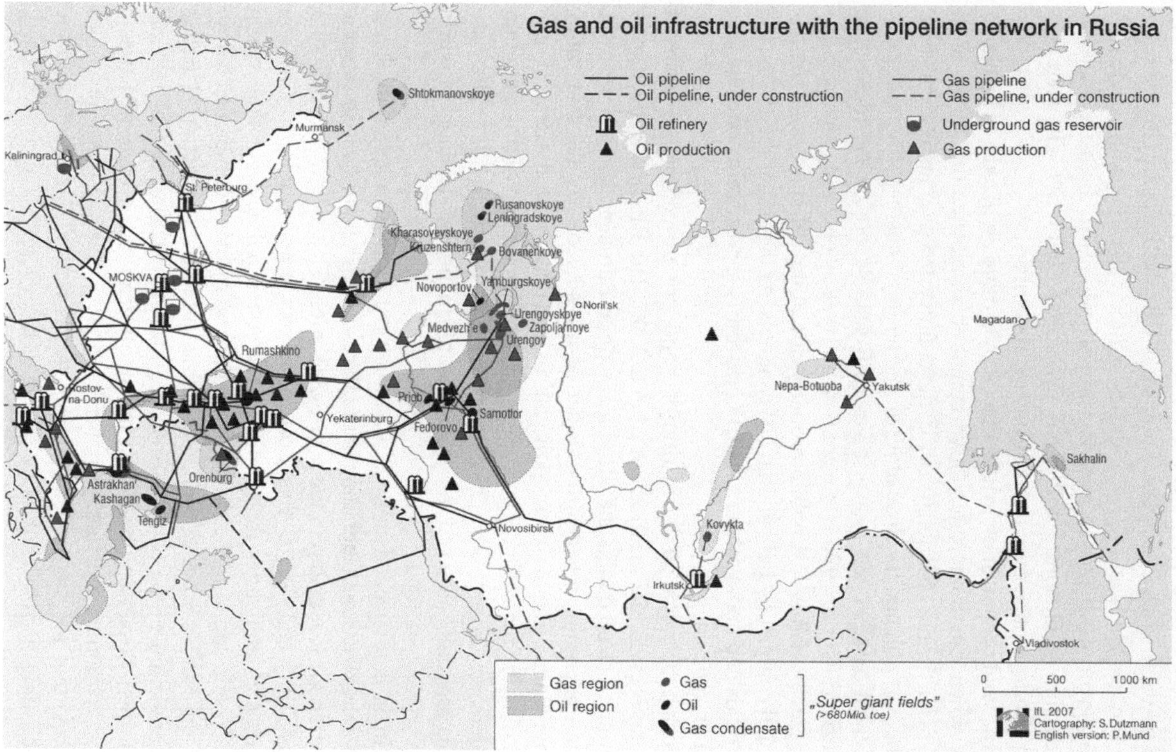

Figure 8.1 Oil and gas reserves and pipelines in Russia

Source: Leibniz-Institute for Regional Geography, Leipzig, 2007.

The privatization process was initiated by Yegor Gaidar's Government in 1991. It led to quick and unfair privatization of the most attractive natural resources of the Soviet Union by governmental officials and managers of energy ministers.[4] The great success of officials in the Ministry of Gas Industry was that almost all gas fields were kept by the 'natural monopoly'. The majority of the monopolist gas company Gazprom's shares went to its managers. Soon after that the Russian Government, headed by the former Minister of Gas Industry Viktor Chernomyrdin, signed an agreement with Gazprom. It passed all government's shares to a trust ownership of the new leadership of Gazprom with Rem Vyahirev as the CEO. The state lost transparent and public control over its most powerful company.

The period of 'wild privatization' was replaced by a more civilized and legalistic course in 1996, when Chernomyrdin's Government accomplished one of the most controversial deals in the post-Soviet history of the country. It was a 'loans-for shares' deal between the Kremlin and its close allies in the national business community. A group of the richest bankers and industrialists in Russia supported the re-election campaign of Boris Yeltsin in 1996 and were paid for that by obtaining shares in the most attractive pieces of Russia's public property, including oil and gas fields, on-ferry metallurgical companies and national TV channels. At the end of the 1990s, the majority of oil fields and about 50 per cent of Gazprom's shares were controlled by private companies and banks.

The relationship between public institutions and private companies changed radically after the presidential elections of March 2000. The Russian leadership in the new millennium has no doubts that Russia is a great European and Asian power that is, however, undergoing a very difficult period in its history. At the same time, neither President Putin, nor his close political allies (Dmitry Medvedev, Sergey Ivanov, and Nikolay Patruchev) have ever argued for the restoration of Russia within Soviet era borders, or for any other form of aggressive power politics. They are confident that Russia does not need this type of confirmation of its status in world politics.

The Kremlin is convinced that Russia has comparative advantages in world politics due to many reasons, including natural resources allowing it to assume economic and strategic leadership in the post-Soviet area. That is why Moscow sees that it has a possibility and right to speak on behalf of the post-Soviet countries in times of most crucial challenges for European and Asian politics. For that reason the Russian leadership supports the activity of public and private companies within the territory of the former Soviet Union, as well as in the EU's direction. There is also a strong conviction in the Kremlin that an inflow of Russian capital into the economies of former Soviet republics will inevitably lead to a growth of Russian influence on the economic and foreign policies of these states. Energy resources are crucially important for the implementation of this policy. For this reason the state is re-establishing its role as the key player in the energy sector of Russia. It is well known among the Russian leadership that only a minor part of the world's oil is produced by private companies and the major part by state-owned companies

4 Joan E. Spero and Jeffrey A. Hart, *The Politics of International Economic Relations* (New York: St. Martin's Press, 1997), p. 342.

occupying monopolistic positions in their own countries. Situation in Russia's oil sector is currently totally different – there are only a few state-owned companies, and with the exception of Rosneft, they are rather small. On this basis the Russian leadership most likely ponders why could the model of huge state-owned oil companies not work properly in Russia, just as it works in Norway, Saudi Arabia, Brazil, Malaysia, and many other countries.

This model, which, however, has by no means been fully adopted in Russia, is not quite perfect. There is no guarantee that the aims of Russian policy in the post-Soviet area will not change radically in the processes of increasing Russian power there. The new priorities of Russia's leadership after the 2007–8 election cycle represent another unknown. And neither is there any guarantee that Russia's western neighbours would not react to its growing power by constructing a new *cordon sanitaire*, which may be based on the amorphous GUAM alliance or the even more illusory project of the Community of Democratic Choice.[5] It is absolutely clear that any abrupt actions of Moscow will immediately lead to counteraction by the US, the most loyal allies of the CIS states in the EU (primarily Poland and the Baltic states) as well as to reactions within the CIS area itself (in Georgia, Moldova and Ukraine). All types of actors in Russian politics would become involved into a conflict of this type. That would make for a disaster in Russia's transformation process. In order to prevent this scenario, Russia, in fact, needs better control over its energy resources. This is also why Moscow has never viewed oil and gas just as an ordinary commodity. They are viewed as Russia's strategic assets employable for foreign policy purposes as well as for bargaining with leading nations of the world on all issues relating to Eurasian economic and security policy.[6]

Re-establishing the state's influence and leading role in the energy sector was a great challenge for Putin and his team. And in 2006, the task was still not quite fully accomplished. Putin quite easily changed the leadership of Gazprom in spring 2001 and consolidated into the hands of its leadership at least half of its assets, which had been trusted to 'friendly companies' by the previous generation of Gazprom's managers.[7] In the next stage, Kremlin staffed the leadership of the country's largest energy companies with prominent figures from the presidential administration: for example the head of the administration Alexander Voloshin, as well as his deputies Dmitry Medvedev and Igor Sechin.

Many commentators of Russian politics are considering the inclusion of the Kremlin's 'commissars' into the boards of energy companies as a soft way of nationalizing private energy companies. Yet, the term nationalization does not fully describe the quality of the relationship between the state and private energy companies during the period of changes in 2002–5. The only fully correct

---

5 The GUAM includes Georgia, Ukraine, Moldova and Azerbaijan. The Community of Democratic Choice was established in Kiev in December 2005 and includes the Baltic states, Georgia, Macedonia, Moldova, Romania, Slovenia and Ukraine as its full members.

6 Amy Myers Jaffe and Robert Manning, 'Russia, Energy and the West', *Survival*, 43/2 (Summer 2001): 133.

7 Peter Rutland, 'Putin and Oilgarchs', in Dale R. Herspring (ed.), *Putin's Russia: Past Imperfect, Future Uncertain* (Lanham: Rowman & Littlefield Publishers, 2005), p. 171.

observation at this point is that the state institutions established control over the making of strategic decisions in the field of transportation of energy. In the oil sector, the only success of the Russian Government apart from the transportation of oil is the transformation of Rosneft into the second largest national oil company – the only state-controlled oil company among the five largest oil companies in Russia. The results in the natural gas sector were more obvious: the presidential administration has established dominant (but not total!) control over exploration of natural gas as well as transportation via Russian territory and the territories of many neighbouring countries. It also remains within the presidential administration's powers to allow entry for transnational energy companies into the Russian market. State institutions are responsible for deciding the routes of new pipelines and locations of new port terminals for exporting oil and oil products.

A geopolitical reasoning relating to the competitive nature of interests and the concomitant wish to establish control offers the first building blocks for an adequate framework for understanding the motives and arguments in the Russian Government's strategic planning. At the same time, we should keep in mind that pure economic factors are also taken into consideration very seriously by the state institutions. As Andrey Gaidamaka from Lukoil has once observed, the total losses of the Russian Federation due to the underdeveloped oil pipeline system make up for about 13 billion US dollars per year. This includes 4 billion US dollars owing to constraints in exploration and export; another 4 billion US dollars represents price discount for monopolistic buyers of Russian oil in Eastern Europe; and about 5 billion US dollars is paid by Russian oil companies for transit via foreign ports and terminals.[8]

There was opposition in Russia's business community to the state's attempts to force private energy companies to abandon their practice of tax optimization. However, they were pushed to pay all taxes in full to the federal and regional budgets. Due to illegal schemes of exporting oil, and registration of oil companies in Russia' internal tax havens, the Russian budget was every year drained of hundreds of billions of roubles. As a result of Putin's success in putting in order the tax payments, the richest people of Russia have lost their status as 'oligarchs', in this context referring to independent actors in national politics and economy equal to the Kremlin and the government of the country. Today these people are simply very rich, but their influence on national foreign policy and foreign economic strategy has declined radically.

The transition period from Yeltsin's deal with the oligarchs to Putin's policy of equal proximity of all types of businesses to state authorities was completed by Summer–Autumn 2003, when several owners of the oil giant Yukos were arrested. Today in Russia, the situation to a large extent is similar to the early 1990s, when the state institutions played the leading role in the energy sector. Now, the main challenge for Russia is the creation of a new system of relationships between the interests of institutions of state power and private companies. Experience tells that decision-making in the thus transformed energy sector should be based on the priorities of national interests, not interests of publicly owned companies mixed

8 A. Vinkov, A. Ivanter A., I. Rubanov and D. Sivakov, 'Tsena bezdejstviya', *Expert*, 38 (11–17 October 2004): 39, 43.

with direct or indirect interests of bureaucrats. Public institutions may well keep the role of strategic decision-making in the transit of energy and regulation of foreign companies' access into the Russian market. But at the same time they should abstain from involvement into the routine management of energy companies. Experience of other countries demonstrates explicitly that private capital is a more effective manager than bureaucrats, and there are only slim chances that functionaries in the Kremlin or in the Russian Government will disprove this assessment.[9]

Energy Policy Priorities in the Putin Era

During the first half of 2000, Putin signed a number of documents defining the major characteristics of Russia's security agenda and its foreign and defence policy. The distinctive feature of all these documents is the priority for economic development as the key element of Russia's security, and the conviction of its role in defining Russia's status in contemporary international politics. Energy sector is not much mentioned in these documents, but the documents do fix a new trend of the economization of national foreign policy, first of all via the protection of Russian companies' interests on foreign markets by the Russian Government and embassies.

At the same time, energy policy problems as components of national foreign economic policy are analyzed in Russia's energy strategy document until 2020. The first version of this strategy was adopted by the Russian Government in 2000. In September 2003 it was enlarged and concretized for the period up to 2020. It is a very detailed and technical document oriented on the growth of energy production. But there are no doubts vis-à-vis another important message in the document: energy sector should be an area of co-operation between institutions of state power and private companies, and one where the Government and Parliament of Russia will play the leading role. During a summit with the then German Chancellor Gerhard Schröder in October 2003, Putin expressed Russia's new approach in this area very directly by saying that 'we are not going to divide Gazprom. The EU Commission should not have any illusions: in the case of gas, it has to deal with the state' (i.e., Russian Federation). The gas pipeline network is the creation of the Soviet Union, and it is only the Russian Federation which can keep it in a functioning order, even if we speak about its parts located outside Russian territory'.[10] This declaration clearly conflicted with the preferences of those politicians in Washington and in European capitals, who at about the same time wanted to open up the Russian energy market for transnational companies which would then establish dominance in Russia's energy sector.

Nowadays about 80 per cent of Russia's oil and gas are exported westwards via European countries or the Bosporus straits.[11] The problem of securing demand can be solved by constructing a new pipeline to Asian markets and by building new terminals with access to open seas. These measures will lead to crucial changes in the number of involved actors in the energy sphere in Russia and to a dramatic increase

---

9 Thomas F. Remington, *Politics in Russia* (New York: Longman, 1999), p. 176.

10 ITAR–TASS online, 9 October 2003.

11 Ivan Rubanov, Neftegazovaja promyshlennost, *Expert*, 37 (4–10 October 2004): 125.

in the number of Russian regions and foreign countries whose interests should be taken into consideration in all stages of the development and implementation of energy policy. In the area of oil exports, the largest projects are in Russia's extreme north (from Surgut oil fields to the Murmansk terminal) and the giant project of transporting eastern Siberian oil to the Pacific coast of Russia and exporting it to customers in Asia and North America.[12] It is hardly possible for Russia to replace Europe as the main destination of Russian exports of natural gas. This is why in the aftermath of the Russian–Ukrainian 'gas war' in 2005–6 (see Chapters 2–4) the priority is to diversify gas pipeline routes bypassing Ukraine's territory. And again, different actors in Russia must be involved for successfully implementing this strategy.

* * * *

The oil and gas industry is largely in private hands in Russia. But links with institutions of state power remain strong. This situation allows for different actors with competing interests in the energy sector: public and private, national and foreign, conservative and liberal, federal and regional, large and small, even patriotic and cosmopolitan. In the remaining parts of this chapter the aim is to cover those actors who can influence the various processes in Russia's energy sector and the EU–Russia relationship in the areas of energy exploration, transit and consumption.

President and Presidential Administration

The current constitution of the Russian Federation was written in autumn 1993, immediately after the short and bloody conflict between President Boris Yeltsin and the then closest equivalent to parliament, the Supreme Soviet. Russia's constitution in the Putin era represents an imprint of Boris Yeltsin's victory over his opponents. This key document reflects the reality of autumn 1993 – Yeltsin's total domination over his opponents and the presidential administration over all other institutions of state power, including the Government, Constitutional Court and both chambers of the parliament.[13] The constitution established a new reality in Russia: the only centre of real power in the country is the president, and any contacts between the president and other state institutions should involve the presidential administration as the only mediator. It is also important to stress that the current role of the presidential administration is close to the one formerly performed by the Politburo of the Central Committee of the Communist Party. The Politburo was the real ruler of the giant superpower although it managed the country via a number of mediators (the government, regional authorities and multiple quasi-parliaments called Soviets). These mediators could be blamed for any failure in their activities, but any political or economic success was directly attributed to the Politburo.

---

12 Andrey Tsunsky, 'Vostochnaya likhoradka', *Expert*, 31 (23–9 August 2004): 24.

13 Archie Brown, 'Vladimir Putin's Leadership in Comparative Perspective', in Cameron Ross (ed.), *Russian Politics under Putin* (Manchester: Manchester University Press, 2004), p. 11.

Putin demonstrated his interest in using natural resources as an important instrument of state power already in the mid–1990s. In 1996, he defended his PhD dissertation in economics in the Mining Institute in St. Petersburg. The topic of his dissertation was 'Strategic Planning of Mineral Resources Reproduction of a Region in the Process of Establishment of Market Economy'. In his work, Putin demonstrated his conviction that public ownership of Russia's primary natural resources has a huge significance for the restoration of the country's economy after the disintegration of the Soviet Union. There is also another conviction in the dissertation: market forces cannot provide favourable economic conditions and social stability for the Russian people in the process of transition from a socialist system to market economy. In order to maintain positive dynamics in the exploration of mineral resources, an important factor is to attract foreign investments into the sector; but even in this scenario, public authorities should retain dominant positions in the sector: 'independently of whose property is natural, including mineral resources, Russia should have legal right to regulate the process of exploration and utilization, defending interests of society in general, and regulating relations between owners of resources…'.[14]

After being elected for presidency in 2000, Putin condemned the chaos during Yeltsin's tenure in 1991–9. Putin coined the idea of the 'axis of power' which connotes the re-establishment of order in Russia's political system and in key sectors of the national economy. Putin devoted his first presidential term to the strengthening of the state's role in energy companies by consolidating the blocks of the government's shares and by changing the taxation system in favour of the state budget. During his second term, which started in May 2005, his clear priority in the energy sector became the creation of large public companies able to function and compete with currently existing private companies. According to some experts, this priority may be explained by preparations of the presidential administration to the 2008 presidential elections. Unlike in the elections of 2000 and 2004, when Putin's win over his competitors was overwhelming, the elections of 1996 demonstrate how electoral results depend on access to the media and financial resources accumulated by the candidates. Russia is set to get a new president in 2008, and that is why the events of 1996 may be repeated. In Russia, energy companies control TV channels and newspapers, and have the financial reserves needed for successful campaign.

A number of changes in the leadership of the presidential administration in autumn 2005 reflected a new tendency in Russia – the growing role of the government as a result of the recruitment of key figures from the presidential administration who consequently brought with them influence in important policy areas. The head of the presidential administration, Dmitry Medvedev, who is a long-term colleague of Vladimir Putin since the turn of the 1990s, became the First Deputy Prime Minister. Sergey Ivanov, a close friend of the Russian President, added to his job as the Minister of Defence responsibilities of Deputy Prime Minister. This means that the two most visible politicians from Putin's close circle assumed strong positions in the government. This gives us an opportunity to predict that in the near future, the

---

14 Putin, Vladimir, 'Mineralno-syrievye resursy v strategii razvitija rossijskoj ekonomiki', *Notes of the Mining Institute*, 144/1 (St. Petersburg: Saint-Petersburg Mining Institute, Technical University, January 1999), p. 6.

role of the President and his administration will shrink compared to the power of the government and its role in the economy. At the same time, it is worth noting that Medvedev's replacement in the presidential administration was Sergey Sobyanin, the former mayor of the oil rich city of Kagalym and governor of the Tiumen region.

The Government and Ministries

As mentioned above, the role of the Russian Government in the management of the national economy started growing after autumn 2005. At the same time the President remains responsible for the strategic development of Russia's economy and regulation of macro-economic instruments (taxation, money supply, exchange rate, etc.). But there are many actors to be taken into account in the complex and still non-transparent system of the executive power in Russia.

A logical starting point is the *prime minister*. Analysis of the activities of all eight prime ministers of post-Soviet Russia demonstrates that the phenomenon of a 'technical' prime minister fits well with the current presidential rule in the country. The power of Russia's Government is growing due to the role of two Deputies of Prime Minister Mikhail Fradkov: Sergey Ivanov, and Dmitry Medvedev, whose two very important positions in addition to his role as Fradkov's first deputy include the chairmanship of the Board of Gazprom and responsibility for the so-called 'national projects'. The latter represent a fairly new element of national economic policy, and are designed to spend in an effective manner billions of petrodollars obtained since 2000 with the dramatic increase in oil prices. The national projects are designed to reinforce the positive image of Russia's leadership especially on the eve of the 2007–8 elections. This new task of Dmitry Medvedev is a very attractive one for any statesman. He is personally responsible simply for spending money from the federal budget whilst having no role in the revenues accumulation side, inflation, or any other unavoidably negative aspects of the new programme's implementation for the national economy. For 2006, the budget for the national projects programme in public health, education, agriculture and housing is about 6 billion US dollars.

The Ministry of Economic Development and Trade is responsible for the strategy on developing Russia's energy sector. The ministry is an outgrowth of the 1990s project of constructing a liberal economic order in Russia, something that is well present in the personal ideology of the powerful Minister of Economic Development and Trade German Gref. The mission of the ministry is to develop a general strategy of public policy and legislative regulation of national economic development. The ministry is also home to the Federal Customs Service, a key element in controlling the foreign operations of Russian enterprises, first of all oil and gas companies.

The Ministry of Industry and Energy supervises the energy sector. The Minister, Viktor Khristenko, previously served in several governments both as a Minister and as Deputy Prime Minister. He has extensive contacts with foreign governments, as well as in the Russian business community, developed during his tenure as a Deputy Prime Minister responsible for Russia's energy sector. The ministry is the most important institution in the day-to-day regulation of the national energy sector, especially in the protection of Russian energy companies interests' abroad. For example, the ministry was working hand-in-hand with Gazprom in 2005–6 during the long and controversial negotiations with Ukraine on the conditions for the transit of Russian gas to European customers and the pricing of gas deliveries to the Ukrainian market. The ministry is also partner for the national monopoly Transneft which controls practically all oil pipelines in Russia. All new pipeline projects have to be developed together with the ministry and later officially approved by the minister. The ministry also maintains technological and economic expertise in the energy sector's investment projects, and makes propositions for improving the use of major oil pipelines, pipelines for oil products, and electrical grids. Further, the ministry comprises the Federal Energy Agency which is responsible for using and controlling the federal budget's funds for implementing strategic energy projects – construction of pipelines and terminals, coordination of public companies' collaboration with private companies, and joint ventures between Russian and foreign enterprises.

The Ministry of Natural Resources (Minister Yuri Trutnev) elaborates the state policy and legal control instruments in the sphere of exploration, utilization, reproduction and protection of Russia's natural resources. There are two institutions worth mentioning in the ministry's structure. The first one is the Federal Agency for the utilization of mineral wealth, and which is responsible for the geological research, economic and financial evaluation of new mineral resources, and for the organization of auctions for the right to utilize Russia's mineral wealth. The agency issues licenses for the exploration of minerals and proposes Russia's regions the exact rate of payments into their regional budgets in return to the utilization of their deposits of mineral resources. The second pivotal institution within the ministry is the Federal Service for Surveillance in the use of natural resources. This service, among its other functions, issues state licenses for exploring mineral resources offshore, and for the construction of offshore pipelines in Russia's territorial waters as well as the continental shelf.

The Ministry of Finance, with its very powerful Minister Aleksey Kudrin, exercises indirect but very substantial influence on Russia's energy sector. The ministry comprises the Federal Taxation Service. This service is responsible for the correctness and timing of tax payments by energy companies to federal and regional budgets, as well as for numerous other payments determined by the Russian legislation. The failure to comply with all payment requests may lead to a collapse of companies of any size, as the example of Yukos during 2003–5 demonstrated very clearly.

The Ministry of Foreign Affairs has experienced several ups and downs in its influence during its post-Soviet history. In several occasions it has been blamed by governmental officials for failing to promote Russian energy interests in the Caspian area, Persian Gulf or Central Asia. Consequently, Putin has become personally actively involved in the implementation of foreign policy, especially with regard to the energy relationship with European countries.[15] From this also follows the practically non-existent role of the ministry in the energy sector. Within the Putin regime, the influence of Sergey Prikhodko, advisor to the president for foreign policy affairs, is much bigger than Foreign Minister Sergey Lavrov's. In principle, Russian embassies and consulates abroad should protect the economic interests of Russian companies, but there are only a few such examples. Russia's leading energy companies are actively expanding their activities to practically all parts of the world. But in large-scale projects – for example, Gazprom's involvement in Germany's gas sector, or Lukoil's activities on the eastern coast of the US – they seek the president's personal support and never rely on the Ministry of Foreign Affairs. This makes the ministry's role in the energy sector negligible.

The Parliament

The Parliament of the Russian Federation consists of two chambers, the upper house called the Federation Council (198 members), and the lower house, the state *Duma* (450 members). The real power of the parliament has never been significant. After the 1993 crisis between President Yeltsin and the national parliament, the parliament's role decreased even more. The years 1993–9 were characterized by a non-stop conflict between the president and the state *Duma*. The main areas of contradictions were approval of annual federal budgets, legislation on taxation of energy companies, and the approval of prime ministers (especially in 1997–9). The largest Russian companies had their own MPs voting in the interests of their sponsors.

The situation changed radically with the December 1999 elections. Since then, pro-presidential parties and political movements have enjoyed a comfortable majority in the parliament. Consequently the Kremlin has not faced any opposition in the parliament and has been able to adopt legislation freely, especially with regard to radically reforming the national economy. This was not imaginable during Boris Yeltsin's rule. Since the parliamentary elections of December 2003, pro-presidential parties have enjoyed more than a two thirds' majority in the chamber. This comprises the so-called constitutional majority, allowing for changing the constitution and the adoption of fundamental constitutional laws. Summing up, the Russian Parliament is now a quasi-independent institution totally controlled by the pro-presidential party United Russia. The parliament supports all legislation initiated by the government and presidential administration, and effectively blocks all legislative proposals from the outside.

In both the upper and lower houses, there are several committees influencing the energy sector: five in the upper house and three in the lower, with their responsibilities ranging from regulatory, economic, industrial, and natural resources to transportation issues. Both houses co-operate closely with experts of energy companies, and have an opportunity to communicate directly with the CEOs of Russian oil and gas

15 Katinka Barysch, 'EU–Russia Economic Relations', in Oksana Antonenko and Kathryn Pinnick (eds), *Russia and the European Union* (London: International Institute for Strategic Studies, 2005), p. 124–5.

companies. The practice of parliamentary hearings is getting more and more popular in Russia, and it gives an opportunity for a direct communication between the legislators and the business community. At the same time, it is not any more possible to point at MPs sponsored by Russian companies in the state *Duma* or the Federation Council. The only exception is Gazprom, the energy arm of the Kremlin.

Russia's Regions

Relations between Russia's regions and centre of federal powers in Moscow give us a lot of examples of a non-stop competition:

1. for the powers to use tax proceeds; energy companies are the largest taxpayers in almost each region of northwestern Russia;
2. for funds from the federal budget; in some regions, federal transfers provide up to 85 per cent of the whole regional budget;
3. and most importantly, for the control over natural resources (oil, natural gas, ore, etc.), and over the exploitation of natural advantages such as geographical location on a border with a large trade turnover.

This competition led to the weakening of power structures within the Russian Federation during the 1990s. The worst-case scenario, the disintegration of the whole state, was a very real threat. In the four months after Vladimir Putin's promotion to power by Yeltsin, Putin formulated as the main task for the federal authorities in general and for himself as the president the re-establishment of the vertical axis of power. The idea of the new leadership team in Russia that assumed power during autumn 1999–spring 2001, replacing Boris Yeltsin and the first generation of Russia's home-made democrats and criminalized oligarchs, was to put limits on the uncontrolled economic freedoms of a number of Russian regions. About 60 Russian regions had received economic powers from Yeltsin.[16] This led to the disintegration of the nation's common economic space and to a growing economic and political separatism. Evidence of an increasing pressure from the presidential administration and the Government of Russia towards the regions since 2000 includes:

1. reform of the Federation Council, where the heads of regional executives and legislative institutions lost their seats and are represented by their plenipotentiary representatives, who are equally dependent on the regional leaders, being appointed them, and on the federal authorities, who provide them with everything they need for conducting their duties, including salaries;
2. establishment of a largely unnecessary institution, the State Council, with regional leaders as members, but without any powers except opportunity once

16 The mechanism of obtaining additional powers from the federal authorities was known as 'treaty on delimitation of powers' between a Russian region and federal authorities in Moscow. For example, the 'Treaty on the Delimitation of Powers between Federal Authorities of Russian Federation and Federal Status of the City of St. Petersburg' was signed on 13 June 1996.

in three months to meet with the president in a company of around hundred other officials;
3. redistribution of collected taxes, strengthening the federal authorities' control over revenues from exploration and export of mineral resources;
4. establishment of seven federal districts with plenipotentiary representatives of the Russian President as mediators between the institutions of federal power and the 89 Russian regions. The seven representatives were shortly afterwards appointed as members of the Security Council of Russia, which is the most influential national institution for strategic planning of Russia's economic, military, social and transport security, as well as responsible for drafting the main guidelines for the country's foreign policy.

Rivalry between Russian regions for serving as a transit territory for oil and oil products exports is intense. There is a growing number of examples of sharp disputes between different regions for getting governmental support for the development of new routes for energy transit, with transfers of federal funds and corresponding changes in the taxation regime and budget procedures for the regions under consideration. Here one can mention the dispute between St. Petersburg and the Leningrad oblast for the construction of a number of export terminals, as well as the interest of Pskov and Novgorod regions for maintaining good economic relations with Latvia and Estonia in order to receive proceeds from railway transportations of oil and oil products to the terminals in Ventspils, Tallinn and Klaipeda.

It is important to note that the most successful projects related to Russia's transit infrastructure for oil exports have been conducted in the Baltic Sea region. The problem from the Russian point of view is that oil through Russia's new terminals on the shores of the Baltic Sea goes to world markets via the Danish–Swedish straits. Moscow already has experience of difficult negotiations with Turkey on the transportation of Russian oil through the Bosporus to the Mediterranean. The Russian business community expressed its concern to the national government about the rhetoric of especially the Baltic countries of the claimed special status of the Baltic Sea as a sensitive ecological system. The businesses' fear was that such labels may lead into limitations in the transit of Russian oil and oil products to the western markets (see also Chapter 7).[17] The solution may be found in a rapid construction of new oil terminals on the Arctic coastline of Russia, near Murmansk, as well as in the far eastern part of the country. This will involve more Russian regions in the country's energy policy.

The Northwestern Federal District

St. Petersburg

St. Petersburg is the leading economic centre of northern Europe. This attracts many energy companies to the city to establish petrol stations, distribution

17 A. Vinkov et al., 'Tsena bezdejstvija', p. 40.

centres, representative offices, analytical centres, etc. Very recently Gazprom and Transneft relocated the headquarters of some of their businesses from Moscow to St. Petersburg. This brings more income to the city's budget and increases the role of St. Petersburg and its administration in the growing energy sector co-operation with neighbouring countries and regions of northwestern Russia. For many years, Surgutneftegas occupied the leading role in St. Petersburg due to the proximity of the Kirishi refinery to the city, but since Valentina Matvienko was elected as the new Governor of St. Petersburg in autumn 2003, Lukoil made more headways than any other energy company with its projects of constructing dozens of new petrol stations in the city and with growing sales of oil products.

The administration of St. Petersburg has always monitored closely the construction of oil export terminals on the coast of the Baltic Sea. Despite the obvious fact that the neighbouring Leningrad oblast is much better located for these projects compared to the northern capital of Russia, St. Petersburg's administration made very energetic attempts to include the Federal Government for the building of port facilities in the city's area. In fact, St. Petersburg's administration suggested several smaller terminals as an alternative to larger new sea ports in the Leningrad region. Initially the idea was to receive more funding for the modernization of St. Petersburg's sea port, but thereafter other possible sites were identified close to the settlement of Volodarskij (in the South-western part of St. Petersburg), and in the city of Lomonosov, which is part of St. Petersburg as well. Governor Matvienko is one of the most powerful and influential regional leaders in Russia and also very successful in converting her influence into growing activities of private national and foreign companies in the city. But due to the small size of the city's territory, it is very difficult to imagine St. Petersburg playing a significant role in the transit of oil and oil products, or to serve as a site for a refinery. At the same time, St. Petersburg is the optimal location for communication between Russian and European energy companies, as well as for the headquarters of regional offices of transnational energy companies.

Leningrad Region

The Leningrad region is currently the leading Russian region in the growing energy co-operation with the EU. As a response to the attempts of St. Petersburg in the late 1990s of relocating transit port facilities inside a city with 5 million inhabitants, during 1999–2001, the Government of the Leningrad region successfully lobbied for the construction of new terminals onto its own territory by the Finnish Gulf.

The main infrastructure project in the region is the construction of oil terminals in Primorsk as well as a pipeline from the west Siberian oil fields into the new terminals. In 1993, the Russian Government decided to construct a 2,400km pipeline from the Timan Pechora oil fields in the Komi Republic to a planned port on the Russian shore of the Gulf of Finland. This resulted in about six years of discussions, and serious concerns from Finland and the Baltic states on possible environmental damages ensuing from the plan. Only in 1999 did the construction start and three years later for the first time Russia's northern oil was shipped to European markets.

All details of the project and its implementation are unique to the Russian economy. The final plan for the pipeline to Primorsk and the oil terminals there was developed in close co-operation between Transneft, and another state monopoly, Russian Railways, and leading Russian oil companies (Figure 8.2). Construction was completed with high quality in a very short period of time. A special tax for energy companies was the only source of financing the huge project, and thus it did not burden the state budget of Russia. The project also served Russia's foreign policy interests by putting leverage on the Baltic states, especially Latvia's Ventspils, which lost an important source of transit income. Russia's energy transit dependence on these three countries was cast upside down. The successful functioning of the Primorsk oil terminals resulted in delaying the construction of new terminals on the southern coastline of the Finnish Gulf. For example, the Russo–British oil company TNK–BP decided to freeze construction of its own oil terminal in Vistino Bay near Ust-Luga on the Russian–Estonian border. This decision was partly based on the high costs of the project and the low costs of using the Primorsk facilities, and can partly be explained by the high export duties for crude oil. This makes refining oil and export of oil products economically more attractive.[18]

Figure 8.2 Pipelines and pipeline projects in North-western Russia

Source: General Libraries, University of Texas at Austin.

Another line of defence for the Leningrad region's authorities in the conflict with St. Petersburg's administration was an idea of modernizing the only sea ports that already existed in the region – the sea ports in Vyborg and Vysotsk (120km from

18 Elizaveta Dobkina, 'Porta ne budet', *Expert Severo–Zapad*, 44 (21–7 November, 2005): 6.

St. Petersburg). The ports in Vyborg and Vysotsk have a long tradition dating back into the early years of the Swedish rule over the area.[19] The main specialization of these ports before 2004 was pulpwood and ore with a maximum capacity up to 4.4–4.5 mty.[20] Compared with the sea port of St. Petersburg and the new terminals in Primorsk, ports in Vyborg and Vysotsk are rather small.

The largest oil company in Russia, Lukoil, started construction of port facilities for export of oil and oil products in Vysotsk in June 2002. At that time the Primorsk terminals were functioning successfully, but they were oriented for the export of oil. The only outlets for export of oil products were located in St. Petersburg (5 mty) and on seasonal terminals on the island of Kotlin near Kronshtadt (4.5 mty). The Lukoil project of building oil terminals in Vysotsk included since the very beginning both the construction of terminals for oil and for oil products. The Russian Government supported the construction by assigning publicly owned land and providing technical support from different ministries and governmental agencies. Lukoil promised to hand over some construction projects at Vysotsk's Lukoil II terminal to Russia's Government. Several Finnish, American and Dutch companies were involved in the construction, which cost Lukoil some 100 million US dollars. As said, the original plan of the company was to export via Vysotsk both oil and oil products (15 mty in total). But the costs for oil exports were much higher than via neighbouring Primorsk due to the necessity to carry oil to terminals by railway tank-trucks. [21] Consequently, in mid–2005 Lukoil decided to change the initial plans. TNK–BP was allowed to use these facilities for exporting its own oil.[22] Lukoil's own plans for spring 2006 included exporting up to 12mty of petrol, kerosene and fuel oil via Vysotsk.[23]

In September 2005, the Leningrad region together with the neighbouring Vologda region became involved in the materialization of the largest and most disputed energy project in Russia–EU relations, the Russian–German project for the Nord Stream gas pipeline from northwestern Russia via Gryazovets and Primorsk to the Finnish Gulf, and via the Baltic Sea directly to Germany, bypassing the Baltic states and Poland. The initial plans included the possibility of selling Russian gas further downstream to the Benelux, France and the UK. The total cost of the project is 5.7 billion US dollars, a large part of which is set to be spent in the Leningrad region (see also Chapter 5). There are also plans to construct near Primorsk a large plant producing liquefied natural gas for export by tankers to customers all around the world.

If at the turn of the millennium the Leningrad region was a junior player in the energy policy of Russia compared to neighbouring St. Petersburg, after half a decade the situation is completely different. The region has attracted the attention of all of the largest Russian energy companies due to its favourable geographical location and

19 Alf Brodin, *Ports in Transition in Countries in Transition* (Gothenburg: Centre for European Research, Gothenburg University, Report 8/2000), pp. 72–3.

20 F.F. Rybakov, *Ekonomika Sankt-Peterburga: proshloye, nastoyaschee, budischee* (St. Petersburg: Gidrometeoizdat, 2000), p. 143.

21 Sergey Ageev, 'Lukoil ukhodit v trubu', *Expert – Severo-Zapad*, 33 (5–1 September 2005): 6.

22 Dobkina, 'Porta ne budet', p. 6.

23 Konstantin Rudenko, 'Vyzhivet silnejshij', *Expert – Severo-Zapad*, 40 (24–30 October 2005): 29.

access to the Baltic Sea, but underlying all this is the active and business-friendly politics of the Regional Government and Governor Valery Serdyukov.

Murmansk Region

The Murmansk region is a relatively new actor in the energy sector of Russia's economy. The two main spheres of interest in the region are oil terminals and exploration of natural gas on the continental shelf of the Barents Sea.

In 2002, the five largest Russian oil companies – Lukoil, Yukos, TNK, Sibneft and Surgutneftegas – agreed on constructing jointly an oil pipeline from the western Siberian oil fields near Surgut to Murmansk. Murmansk is a city in the extreme north of Europe with an ice free deepwater port on the Kola Bay (Figure 8.1). The port has an unlimited ocean access. It is the shortest route for Russian oil to the US market, the largest and most attractive domestic market in the world. The 2002 plan for a pipeline project by the largest Russian oil companies emerges from their interest in increasing exports and easing the bottle-neck effect resulting from a lack of domestic export terminals. Russian oil companies were trying to challenge the monopoly of the fully state-owned Transneft, which has monopolized all strategic pipelines constructed during the Soviet era.[24] The idea was to construct the first private oil pipeline and destroy Transneft's monopoly. Both foreign media and experts of leading Washington-based think tanks (the Carnegie Endowment and the Centre for Strategic and International Studies, CSIS) together with several large transnational oil companies aggressively supported the project. At some point the discussion turned into a crusade of western experts and journalists together with Russian oil barons against the strengthening involvement of the Russian Government in the energy sector.

As we know now, the presidential administration won the conflict and thus maintained its strategic grip over the energy sector, at least up to the next presidential elections in 2008. But the real victim of the dispute was the Murmansk region. A very rational and attractive project was postponed, and the development of the city's port slowed down. There are plans of using railways for transporting oil and oil products from the Komi Republic's oil fields to Murmansk and load these to deep water tankers. But the cost of transporting oil in this way is much higher and less attractive for oil companies. For many decades, Murmansk was a Soviet/Russian fortress in the Arctic next to the NATO member state Norway. The military's role in the region remains very important, making the capital's strategic concerns pivotal in all discussions about the region's economic development.

The Government of the Murmansk region expects to profit together with the neighbouring Archangelsk region from the off-shore oil and gas exploration in the Barents Sea. But again, the role of the regions here is minor compared to the role of the Federal Government and the gas giant Gazprom.

24 Barysch, 'EU–Russia Economic Relations', p. 126.

The Komi Republic

The Komi Republic is one of the richest Russian regions thanks to its huge deposits of mineral resources. The population of the republic is about 1 million inhabitants, and there are oil, natural gas, coal, bauxites and forest resources. The oil fields are the most attractive sector of the regional economy.

Extraction of oil started in the republic in the Soviet era. Since the early 1990s, production of oil was falling every year. Two local oil companies, KomiTEK and Komineft, were privatized by the regional elites in the early stages of privatization. But the activities of the local owners and managers produced disastrous results. Production of oil decreased from 15 mty in 1990 to 3.5 mty by 1998, and budget revenues fell down significantly. Technological degradation resulted in the spoiling of hundreds of hectares of land, and the rivers Pechora and Usa.

The bankruptcy of KomiTEK and Komineft, and Lukoil's subsequent invasion into the republic kick-started the revival of the oil sector within the regional economy. Lukoil bought the bankrupted companies' assets and invested heavily into the extraction of oil (600 million US dollars during the first two years of its activities in the region). As a result, currently about 80 per cent of the Komi oil is Lukoil's production.[25] The company invested also in the modernization of the oil pipeline Harjaga–Usa and into environmental protection in the region. The Komi Republic is one of the main suppliers of oil to terminals in Primorsk and to European markets. In autumn 2005, Gazprom and Lukoil signed quite a unique agreement on strategic co-operation for the period 2005–14 in different Russian regions. One of the regions, which has primary importance for both companies and mentioned in the agreement is the Timan-Pechora oil province in the Komi Republic.

Archangelsk Region

The Archangelsk region is located on different transportation routes of energy resources from Siberia and the Komi Republic to the Baltic Sea and, in the future, to the Barents Sea. Its importance to national energy policy in transit questions is pivotal and increasing.

During two consecutive terms as the Governor of the region (1996–2004) Anatoly Efremov created an efficient system of co-operation between the regional authorities and managers of Lukoil.[26] The region was considered a 'Lukoil area' by other energy companies. Transformation of this system started in 2004, when Anatoly Efremov lost his job and the new Governor Nikolay Kiselev decided to diversify the region's partners in the energy sector. The largest state-owned oil company Rosneft strengthened its positions in the region after the completion of several oil terminals in Arkhangelsk's sea port (4 mty) and in the south of the region near the railway station Privodino (6 mty). Rosneft placed an order to the local naval

25 Aleksey Bessudnov, 'Pospeshaj medlenno', *Expert Severo-Zapad*, 35 (19–25 September 2005): 36, 38.

26 Nadezhda Breshkovskaya, 'Zamknutyj mir', *Expert – Severo–Zapad*, 48 (20–27 December 2004).

shipyard Sevmashpredpriyatie for the construction of oil platforms to be used at the off-shore oil field Prirazlomnje in the Barents Sea.[27] This makes Rosneft the largest local employer and provider of stability and economic development to the region. As will be described below, Rosneft is currently the 'beloved child' of Kremlin officials, providing both the company and its partners huge advantages in economic activities and political power in the country. This also promotes the Archangelsk region into a very special status in Russia's energy policy.

Another important player with growing interests in the region is Gazprom. It is constructing a gas pipeline Njuksenitsa–Archangelsk (350 kilometres) as well as a grid of pipelines around the region. One of these smaller pipelines is especially important for Russia's security, as it will provide natural gas to the largest space centre on Russian soil near Plesetsk and Mirnyi.[28]

Kaliningrad Region

For several decades after the Second World War, Kaliningrad region's development was defined by its special status as the Soviet Union's western military outpost. The navy in particular established itself there, with the Baltiysk base as the home port for the Soviet Baltic Fleet, along with a multitude of other branches of the defence. The Kaliningrad oblast's present position as Russia's exclave in the west, bordering NATO and EU countries, highlights its continuing geopolitical importance.

In early March 2004, Lukoil opened an offshore oil-drilling platform at the Kravtsovskoje oil field. By the standards of Russian oil companies, this production site is fairly small: geological reserves of the deposit are approximately 21.5 million tonnes, and only 9.1 million tonnes can be extracted by currently available technologies. But the project has a high foreign policy profile for Russia and for EU–Russia relations. President Putin personally expressed Russia's interest in the successful functioning of the platform. From the point of view of Russia's interests, it is extremely important to wave the national flag in such an important region as the Baltic Sea area. During the 1990s, this region was considered by Moscow the most peaceful and stable of all regions around Russia's borders in Europe and Asia. By 2004, with Poland, Estonia, Latvia and Lithuania being members of the EU and NATO, the situation had changed a lot. This had made the relations between Russia and its closest western neighbours a non-stop set of conflicts, hostile declarations, and political and even military demonstrations.

For outsiders, the reasons for the continuing Russo–Baltic conflict may not necessarily look very fundamental: disagreement of Soviet–Baltic relations during 1939–91, lack of border agreements between Russia and Latvia, as well as Russia and Estonia, and Russian perceptions of unequal treatment of Russian-speaking peoples in Latvia and Estonia (cf. Chapter 7). But in the geo-strategic sense, a more fundamental challenge for Russia is the growing influence of NATO as a

---

27 Andrey Subbotin, 'Dva napravleniya odnogo puti', *Expert – Severo–Zapad*, 41 (31 October 2005): 32.

28 Interview with the Governor of Arkhangelsk region Nikolay Kisevel, *Expert – Severo-Zapad*, 41 (31 October – 6 November 2005): 26.

military organization on Russia's western borders. Because of Lukoil's activities, Russia wants to demonstrate its physical presence in the region. Lukoil's activity in the region is important also from the economic point of view. The company is constructing terminals for shipping oil to local customers and abroad, and it also built an advanced steel products factory in the town Izhevskoye.[29] This factory produces oil drilling platforms which will be used in the Baltic Sea as well as in the Caspian Sea, and offshore of Sakhalin. Possible partners for the factory are the Dutch company HBG Steel and German NOELL, a subsidiary of the Preussag company.

Pskov Region

This region is probably the least influential in the energy sector among the regions of northwestern Russia. There are no oil or gas fields, or refineries in the region. At the same time, the region is second in northwestern Russia in the transit of goods abroad, handling in 2004 about 68 mty, mostly by railways. A significant share of these goods is oil and oil products on their way to sea terminals in the Baltic states and further to EU markets. The region has suffered from Transneft's decision to switch oil export from the Ventspils terminal to Primorsk. The Pskov region is specially interested in friendly relations with the EU and benefited from the EU memberships of Latvia, Lithuania and Estonia.

Russian Oil and Gas Companies

The influence of leading Russian energy companies on the decision-making in Russia's economic and foreign policy is difficult to estimate. On the one hand, Putin's attempts to decrease the influence of 'oil barons' upon the Federal Government and Parliament were fairly successful. This reduced the lobbying capacity of private energy companies considerably. But on the other hand, economic power and financial resources of Russian energy companies have been growing steadily since 1999. In 2005, according to the *Petroleum Intelligence Weekly*, ten Russian companies were within the top-50 on the list of world's 100 largest oil companies.[30] The Russian Government has a controlling stake only in two of the ten largest oil companies in the country. At the same time, due to the existing legislation and rules of the game in the sector, the Federal Government maintains an option to effectively influence the actions of any companies working in the country.

There are several challenges for Russian energy companies where they need close co-operation with the Federal Government. The first one is related to the underdevelopment of the export infrastructure and its resultant bottle-neck effects for Russia's oil exports, especially with regard to terminal facilities in Russian ports. As one of the consequences of the dissolution of the Soviet Union, foreign trade between the Russian Federation and its major western economic partners had to

---

29 Andrey Vinkov, 'Po zapovedjam Forda', *Expert*, 10 (15–21 March): 28.

30 'The Energy Intelligence Top 100: Ranking the World's Oil Companies', *Petroleum Intelligence Weekly*, special annual edition (2005).

adapt to new geopolitical realities.[31] The most challenging task for the government and the then emerging private oil companies was to find new, financially efficient routes for water-borne transport.[32] The second challenge is related to the lack of domestic investments and the slow speed of taking into use new deposits of energy.

In order to respond to these challenges, private companies not only need to co-operate with the Federal Government, but also with regional authorities. Foreign banks and companies may assist Russian companies in their search for investments and advanced technologies,[33] but again, only the support of the government or presidential administration will create conditions for success. All leading energy companies in Russia are active supporters of full freedom of foreign investments into the Russian economy, the only dispute being the problem of tax exemptions for foreign companies equalling those granted to Russian companies. In the Russian energy sector, foreign policy interests and the logic of entrepreneurship go hand in hand.

In April 2001, Putin invited to the Kremlin the CEOs of Russia's largest oil companies in an attempt to urge them to act more proactively in the strategically important regions on the southern borders of Russia, in the Caspian Sea area. The Russian profile in the region is heavily determined by the long-term and bloody war in Chechnya. Invitation for Russian companies to invest into the region, especially into the oil fields in Azerbaijan and Kazakhstan, as well as into offshore oil fields in Russia's sector of the Caspian, was an attempt to change the image of the country and get a stake in one of the world's most promising oil provinces. Soon afterwards Putin visited the oil drilling platform constructed by Lukoil off-shore in the Russian sector of the Caspian. Similar mobilization of oil companies for the promotion of Russia's political and economic interests can also be expected in any other region of the world.

Gazprom

Gazprom occupies a unique position in Russian politics and economy. Some experts called the company 'Russia's Ministry of Foreign Affairs for the 21st century'. This is true at least in the case of the post-Soviet countries where the company's officials are as powerful as Russian governmental officials and diplomats. Since the disintegration of the powerful Ministry of Gas Industry with the Soviet Union's collapse, Gazprom became used to having governmental support behind it. Viktor Chernomyrdin, the first Prime Minister of Russia approved by a parliamentary majority in autumn 1992, founded the company. During the last years of his tenure as Prime Minister, the only

---

31 Andrei P. Tsygankov, *Pathways after Empire: National Identity and Foreign Economic Policy in the Post-Soviet World* (Lanham: Rowman & Littlefield Publishers, 2001), p. 32.

32 Alf Brodin, 'Breaking Loose? Russian Reasons to End Its Dependence on Foreign Ports', in Lars Hedegaard and Bjarne Lindström (eds), *The NEBI Yearbook 2001/2002: North European and Baltic Sea Integration* (Berlin: Springer Verlag and Nordegio, 2003), p. 123.

33 Isakova, *Russian Governance in the Twenty-First Century*, p. 27.

reason for his foreign trips was the protection or promotion of Gazprom's interests.[34] No one other Russian company had the same influence on national politics and economy than Gazprom under Rem Viakhirev as the CEO and Viktor Chernomyrdin as Prime Minister. But the influence and power of Gazprom did not decrease after the resignation of Chernomyrdin in March 1998 and the 'retirement' of Rem Viakhirev in May 2001. Aleksey Miller, official of the St. Petersburg administration and friend of Vladimir Putin, was appointed as the new CEO. Active support of Vladimir Putin is the key element to the unique stability and growing power of Gazprom since 2001. Despite the tremendous pressure from the US and the EU, Putin has rejected the proposals to divide Gazprom into several smaller companies, whose specialization could be extraction of gas in several major gas fields and transportation through private grids of pipelines. Both gas fields and the whole grid of pipelines continue to rest under the direct control of Gazprom.

Analysis of recent foreign policy initiatives demonstrates that Gazprom is a powerful instrument of Russian foreign policy towards neighbouring countries, now when Russia's leaders have finally been able to formulate national priorities in relation to them. During 2004–5, the Russian Government successfully concentrated about 51 per cent of Gazprom's shares under its control and opened the market for the rest of Gazprom's shares for foreign investors. Largest energy companies in Europe (German's BASF and E.ON, Italian ENI, etc.) are working closely with Gazprom in such giant projects as the Blue Stream and Nord Stream pipelines. President Putin meets Gazprom's CEO more often than the majority of ministers in the Russian Government, and all major TV channels and newspapers cover these meetings as a major event of the day in the country. The same is true for the visits of Aleksey Miller to Russian regions and foreign countries. Gazprom is Russia's largest earner of hard currency and its tax payments account for about one quarter of federal revenues.[35] At the same time, the company's near monopoly position in the exploration and transportation of natural gas limits the industry's development. Restricted access to the pipeline system for independent producers of gas, for example, has resulted in oil companies flaring their gas or selling it to Gazprom on very low prices.

Today Gazprom is very active in the regions of northwestern Russia. It is planning to invest in St. Petersburg several billion US dollars in the coming years for the reconstruction of local heating stations and gas grids in central historical districts of the city. In addition, Gazprom bought the leading local football club Zenit and started constructing a new stadium with total costs of about 300 million US dollars. The investments into gas supply infrastructure and the football team are important indicators of the company's growing involvement in the city's economic activities and social life. In general, there is no governor in Russia who could afford conflictual relations with Gazprom. Gazprom has no critics among politicians, business associations, TV channels, or newspapers. Gazprom started the new century

34 Michael McFaul, Nikolai Petrov and Andrei Ryabov, *Between Dictatorship and Democracy: Russian Post-Communist Political Reform* (Washington DC: Carnegie Endowment for International Peace, 2004), p. 324.

35 Debra Johnson, 'EU–Russia Energy Links', Debra Johnson and Paul Robinson (eds), *Perspectives on EU–Russia relations* (London and New York: Routledge, 2005), p. 186.

as something akin to Russia's Ministry for Foreign Affairs. And its transformation into a parallel government of the country, key engine of its economic growth, and the most important actor in national and European energy markets only took a few years.

Lukoil

Lukoil is Russia's largest oil company. Its shareholders include the Russian Government, management of the company, Russian banks and even the huge transnational company ConocoPhillips. Lukoil has very strong positions in European countries and in the Caspian area. In 2003, Lukoil bought from Getty and Mobil more than two thousand petrol stations on the eastern coast of the US, whilst there are only 1711 Lukoil petrol stations in Russia. The Russian company is supplying its own oil products for retail via a chain of petrol stations abroad, and today Lukoil's sales in the US market alone are bigger than in the domestic Russian market. In September 2003, President Putin personally took part in the opening ceremony of the first petrol station in the US under Lukoil's brand. The CEO of the company, Vagit Alekperov plays a very active role in the Russian–American energy dialogue initiated by Putin and the US President George W. Bush in 2002.[36] Lukoil is also very active in the Persian Gulf, especially in southern Saudi Arabia, where it is planning to pump natural gas and maintains hope of getting an opportunity to make investments into the tremendous Western Kurna oil fields in Iraq. Lukoil views European markets as a natural destination for its oil products produced in Russia, as well as an attractive area for investments into refineries and oil pipelines. As discussed, the company is very active in some regions of northwestern Russia, especially in St. Petersburg and the Archangelsk region.

Rosneft

Rosneft and Zarubezhneft are fully state owned oil companies. This gives them a very distinctive status in the country's energy sector. They control several oil rich fields both in Russia and overseas.[37] For these reasons their activities have both economic and foreign policy dimensions. Up until the early years of the new millennium, there were many representatives of an older generation of managers in the directorates of the two companies without skills of navigating in market economy conditions. The new leadership of Rosneft and the active support of the company's activities by Igor Sechin, the powerful Deputy Head of Putin's presidential administration, brought radical changes. Today both Rosneft and Zarubezhneft are just as successful and efficient as the majority of Russia's private companies.

Due to the support from Kremlin's *siloviki* group, which includes former officers of the intelligence services KGB and FSB now occupying key positions in the

36 Alexander Koksharov, 'Deklaratsija vzaimozavisimosti', *Expert*, 9 (7–13 March 2005): 45.

37 Rosneft and Zarubezhneft control shares of enterprises in Kazakhstan, Ukraine, Algeria, Columbia, Vietnam, and Syria.

presidential administration and some 'power ministries' of the Russian Government, Rosneft has extremely favourable conditions for its business compared to any other Russian company.[38] For example, in September 2004 the company received from the Ministry of Natural Resources ten licenses for oil fields with about 700 million tonnes of approved oil reserves. These oil fields are nearby the Vankorskoye oil field in the Krasnoyarsk kraj that already belongs to Rosneft. This gift from the Russian Government has strategic consequences for Russia's oil industry, because in the mid-term eastern Siberia will replace the current key oil province of Russia, the Tiumen region in western Siberia. Rosneft has replaced Yukos, founded by the now imprisoned Mikhail Khodorkovsky, as the leading Russian oil company in eastern Siberia. The main part of Yukos, Yuganskneftegas, fell into Rosneft's arms after the very controversial bankruptcy of the company in December 2004. By this manoeuvre the Russian Government regained initiative in several strategic projects in the region, including public investments into infrastructure and construction of the disputed pipeline either to Dutsin in north-eastern China, or to Nakhodka in the far eastern coast, with a possibility to export oil to Japan and the US.

It is easy to foresee Rosneft becoming a key actor in Russia's oil sector, as well as an ambassador of the Russian oil industry abroad. The aim of the Rosneft CEO Sergey Bogdanchikov is to make the company in the oil market akin to what Gazprom is in the gas market, i.e. a monopolistic actor in the national market and with an aggressive strategy on foreign markets, supported by the Russian Government, and, in extreme cases, personally by the president.

Surgutneftegas

Surgutneftegas had for many years a fairly strong position on the markets of northwestern Russia. The biggest refinery of the region in Kirishi (in the Leningrad region), is the leading producer of petrol and heating oil in the region. Surgutneftegas also constructs its own terminals on the coast of the Finnish Gulf at the Batareinaya Bay near the Sosnovyi Bor nuclear power station westward from St. Petersburg. The company enjoys the support of the Leningrad region's Government, but its positions in the St. Petersburg area are getting weaker due to the competition with Lukoil.

Surgutneftegas exported significant volumes of oil via Estonian terminals in Tallinn. But since the construction of Russia's Baltic Pipeline System was started in 1999, the company lent its support for the project and switched its oil exports to Primorsk as soon as the terminal was opened there in 2002 (Figure 8.1). Surgutneftegas has always been loyal to governmental policy and probably will keep a low profile in its business strategy both in Russia and in neighbouring countries.

The Mikhail Khodorkovsky court case during 2003–5 demonstrated clearly how in Russia, oil companies lack the necessary economic and political power for getting involved into a conflict with the presidential administration and the Government of Russia. It is not enough to have an effective business strategy, support of foreign financial institutions, other leading companies and the media for influencing

38 For example, the Head of the Federal Agency on Energy Sergey Oganesyan previously held a high position in Sakhalinneftegaz, and was also the vice-president of Rosneft.

the decision-making process in the energy sector. Even the richest and the most successful Russian oil companies remain unable of converting their economic power into a corresponding political influence. This can be explained by the simple fact that Russia's political system is still in the process of formation and that many traditions and rules of the game are non-existent.

Financial Institutions

Financial institutions do not have a long history in Russia and are still weak compared to Western Europe and even the former socialist countries of Eastern Europe. Due to the financial instability in the 1990s, several banking crises and the protectionist policy, the banking sector is still undeveloped and not capable of performing many of the normal functions of a modern banking system, including accumulation of investments for developing the national economy. The system is dominated by huge state-owned banks, Sberbank and Vneshtorgbank. The largest private bank in Russia is the Alpha Bank with total sales of financial services at about 800 million US dollars in 2004. In 2005, there was not a single bank from the northwestern region of Russia in the list of 400 largest companies of the country! Russia has an extremely weak system of insurance companies. They are only successful in closing the domestic market from foreign competition, but very weak in performing the pivotal functions for attracting investments into the energy sector of Russia's economy.

Many Russian energy and industrial companies have their own banks serving their operations and taking care of their financial accounts. But neither are they performing the key functions expected from a modern banking system, and their reserves are very limited for the development of new projects and ambitious investments. This leads to the conclusion that on the whole, Russia's financial institutions are not able to serve the interests of national energy companies within the growing Russian economy. This is a severe hurdle in the modernization of Russia's economy.

The Stock Exchange

Since Russia's economic collapse in 1998, the country's stock market became one of the most dynamic ones in the world, with non-stop growth from 1999 onwards well into the new millennium in all major segments. The share of energy companies within the market is absolutely dominant. Due to high oil prices and improving quality of management in the sector, it becomes more and more attractive and transparent for both direct and portfolio investors from Europe and North America.

The release of Gazprom's shares for stock market trade was for long a key problem in the national financial market. As explained above, in 2005, the Russian Government finally assumed control of over 51 per cent of Gazprom's shares. In 2006, the rest of Gazprom's shares were opened for any investors both from inside Russia and abroad in place of the previous national trade in St. Petersburg. The shares became available also in Moscow, and in London and New York through depository receipt schemes. Gazprom is among the top ten companies in the world in terms of market capitalization, and has quite good prospects of becoming the largest one in the world.

Business Organizations in Russia

Vladimir Putin's election as the President of Russia in March 2000 marked a fundamental change in the status and role of organizations of businessmen in the domestic politics and Russia's economy. During the 1990s, their influence on the economic decision-making in Russia was obvious and quite strong, making the business community a relatively independent centre of power. At the same time, internal competition and fight for influence on Yeltsin and his 'family' divided leading businessmen in the mid–1990s. The only crucial exception to this rule was in January 1996, when seven leading businessmen and group of politicians, headed by Anatoly Chubais, created a temporal alliance for the re-election of the very unpopular Yeltsin for a second term. He was re-elected in July 1996 in a clear demonstration of the then growing power of Russia's wealthiest people known thereafter as 'oligarchs'.

After Russia's financial crisis of August 1998, the aggregate economic power of the 25 richest Russian businessmen significantly exceeded that of the Russian state.[39] The situation was unacceptable for Putin as the new President of Russia, elected in March 2000 without any significant financial assistance from the oligarchs.[40] He was able to mobilize popular support in the wake of a new war in Chechnya and the fairly good economic situation in late 1999 – early 2000. These factors helped him to change fundamentally the relationship between the business community and governmental institutions, without even launching any real dialogue with the former, or recognizing the previous 'results' in their mutual relations. The resulting situation was a totally new one: the business community became expected to recognize the supremacy of Kremlin and the Government of Russia in economic and political life, and negotiate with them on any issues seen as having impact on Russia's stability and economy.

Putin has explicitly refused to differentiate between large and small enterprises when it comes to access to governmental decision-making and law-making processes in the parliament. In order to point the appropriate space of action for the oligarchs he upgraded the status of the Russian Union of Industrialists and Entrepreneurs (RSPP).[41] This organization was originally established as a union of so called 'red directors', i.e. Soviet-era and Soviet-style managers who had successfully maintained control over their enterprises. During Yeltsin's era, they asked and received governmental protection and direct financial assistance. However, Putin transformed the organization into a club of loyal oligarchs and CEOs of the most important companies of Russia, first of all in the energy sector. The new nickname of the organization became the 'trade-union of oligarchs'.

Putin also invented a new tradition of regular meetings with the leadership of RSPP where both sides can speak openly about their concerns and interests, but the results of which were agreed to remain behind-the-doors information. One of these meetings was marked by a conflict with the then Yukos CEO Mikhail Khodorkovsky.

---

39 Forbes World's Richest People List for 2004.

40 Rutland, 'Putin and Oilgarchs', pp. 166–7.

41 William Tompson, 'The Russian Economy under Vladimir Putin', in Cameron Ross (ed.), *Russian Politics under Putin* (Manchester: Manchester University Press, 2004), p. 129.

Two years afterwards the company was effectively bankrupted, and Khodorkovsky himself serving an eight years prison sentence for financial crimes. The Yukos lesson is a very clear one for all other leading businessmen in the country. It led them to lose interest in competing for political power with the Kremlin, and in fact started to use any opportunity to demonstrate their loyalty to the government and the presidential administration. As said, before the 2003 state *Duma* elections, practically all of Russia's leading energy companies had their own MPs, but since then it has become a somewhat risky business for any politician to be known as a particular oil company's representative. This sums up why the role of business organizations on decision-making processes in Russia may be estimated as small.

Foreign Energy Companies in Russia

Foreign companies have been concerned about the development of their Russian business ever since the Russian market was opened for them in the early 1990s. Their primary concern initially was to establish control over the newly founded private oil and gas companies of Russia, as well as to invest into the exploration of energy via production-sharing agreements (PSAs).

During Yeltsin's era, foreign energy companies considered investment climate in Russia as unfavourable. The personal involvement of Putin in the merger of the large Tiumen Oil Company and British Petroleum in 2003 into TNK–BP encouraged foreign companies to expect good prospects of entering the Russian market during Putin's second presidential term. But at the same time, Russian leaders sent a conflicting message by blocking the merger between Yukos and ChevronTexaco, or ExxonMobil. The terms of exploration in the Sakhalin 3 project were also altered, with negative consequences for American investors. This was a clear message that foreign companies should contact Kremlin and the Russian Government directly in order to define the terms of their investment into Russia's energy sector. But in 2004, Putin personally blessed the acquisition by ConocoPhillips of a controlling stake in Russia's largest oil company Lukoil. There are still huge interests by American and Indian companies in the oil fields off-shore of Sakhalin.

Conclusion

In the 1990s, Russia's new political elite concluded that dependence on transit countries and on consumers of Russian energy resources in Europe is dangerous for national security. Russian energy companies complained of high tariffs for transportation and of low quality of the port facilities in the Baltic states. Regional authorities in Russia's European part supported these voices actively, lobbying their interests in the Kremlin, the State Duma and in the national media. But the weakness of Russia in the 1990s, and its inclination to continue neglecting the development of its transit infrastructure under pressure from Western countries led to an economic decline in its energy sector due to a lack of investments and export facilities of its own. Only in 1999 did the energy sector become the highest priority for the Russian Government. Vladimir Putin entered the Kremlin with a vision of the energy sector as a powerful resource for accumulating national power and of overcoming transit limitations that in his view posed a complex geopolitical threat to the future of Russia. The consensus on prioritizing this problem definition emerged quite soon, and since then it has stayed the key element guiding decision-making in the energy sector.

It is thus fair to say that the strengthening of public control over the energy sector – vis-à-vis exploration and transit in particular – is the latest trend in the national economy, started in 1999 and reinforced in 2003 by the cutting of oil supplies to Ventpils and blocking the merger of Yukos with foreign companies. Successful implementation of Russia's plans to develop its energy export infrastructure as an alternative to port facilities in foreign countries and pipelines through the Baltic states pushed the Kremlin to developing even more projects for minimizing its transit dependence on foreign countries. Huge inflow of petrodollars into the national economy has decreased Russia's dependence on foreign investments into exploration, procession and transportation of oil and natural gas. At the same time the experience accumulated during Russia's post-Soviet years demonstrates how any project in the energy sector can only succeed if it balances public and corporate interests. Yet, it is fairly difficult to imagine that anybody in the Kremlin would want to see huge energy companies managed by bureaucrats, when professional managers will clearly be more successful and effective. It can be concluded that accepting these premises, the challenge for the Russian regime is to establish a system of checks and balances in the sector, including means and tools to influence and even press leading energy companies if they fail to support the official line. But clearly, it is in Russia's interests to use these instruments only in the most extreme cases, and in issues such as making decisions on routes of oil and gas pipelines and locations of port terminals in Russia.

The search for a proper formula of public–private partnership is the most difficult task for public policy within the energy sector. All actors analyzed in this chapter are involved in this effort, even if some of them with more ownership of the agenda than some others. As George Orwell once wrote: 'all animals are equal, but some animals are more equal than others'. This more equal actor in Russia's energy sector is the president of the country and his administration, as well as business structures close to officials in the Kremlin.

Today it is obvious that Russia is trying to use its energy resources for improving its standing in European and world politics. The latest policy changes in the sector are made in order to develop Russia into a leading energy superpower. This will mean that Kremlin will allow foreign companies to invest into the energy sector and co-operate closely with leading Russian oil and gas companies only whilst simultaneously advancing its own foreign policy objectives. The interest of Dmitry Medvedev and some of his colleagues for an alliance between Gazprom and Rosneft represents one effort of optimizing the strategic grip of public institutions. The establishment of such a super energy company would make it the world's largest in terms of its key characteristics like total production of oil and gas, and the size of pipeline network. But also the needs for financial resources for the company's modernization and geological research would be huge. However, this idea was effectively blocked in mid-2005 by other politicians around Putin who are competing with Medvedev. The idea of a super energy corporation was forgotten at least for a while.

Dmitry Medvedev's move from the presidential administration to the government should be interpreted as Russia's attempt to create a more transparent system of management in the energy sector and to avoid overlaps between the government and presidential administration in its management. At the same time, it is very possible that prior to the presidential elections in March 2008, the role of the government in national politics and economy will increase radically and the role of the presidential administration decrease. But in either case it can be agreed with the remarks of Martha Olcott in a paper published by the Moscow Carnegie Center: '...supremacy of the Russian state in the energy sector can not be a matter of bargaining'.[42]

Neither the president nor the government oppose foreign capital's presence in Russia. But they argue for a 'managed' and slow inflow of these companies into the domestic market. Strengthening Russia's role in the world's energy market is their top national priority. This is also why Russia placed energy security as a priority item onto the agenda of the G8 summit held in St. Petersburg in July 2006.

Finally, any 'OPECization' of Russia's energy sector should be considered very bad news for everyone concerned. The Russian President is against this scenario of locking the national economy into a third world model. At the same time, the Government of Russia is against any sort of supranational control over the energy sector. Russia's suggestion in 2002 to Turkmenistan, Kazakhstan, and Uzbekistan of a Eurasian 'gas OPEC' should be read in this light, and particularly as something directed against the supranational model for the energy sector's regulation attempted through the EU-led Energy Charter Treaty. This leaves us to conclude that Russia promotes a geopolitical approach to the political economy of energy supply and that this discrepancy of the Russian and EU models allows for speaking of important elements of a zero-sum game in Eurasian energy politics.

42 Martha Brill Olcott, 'Vladimir Putin i neftyanaya politika Rossii', *Carnegie Center Moscow Working Papers*, 1 (2005), p. 5.

Conclusion: Prospects for a Pan-European Energy Policy

Pami Aalto

Introduction

The aim in the preceding chapters was to cast new light on the EU–Russia energy dialogue by deriving from several theoretical approaches. These include the relatively well-established energy diplomacy approach, the geopolitics and energy security approach, energy economics and trade approach, and the energy and the environment approach. In addition, we looked at the energy dialogue's wider ramifications. These include its significance for European integration, EU–Russia relations and Russia's economic and political transformation, as suggested by the political sociology of energy approach. Specific features extending to bureaucratic politics of energy were also covered in northern European EU member countries like Germany, and the Nordic and Baltic states, as well as Russia. And as suggested by the regional politics of energy approach, a further aim was to probe into the federal and regional level struggles in Russia.

The purpose in this concluding chapter is first to summarize what the diverse approaches that we have used as our reference points tell us about issues like who acts in the EU–Russia energy dialogue, or in EU–Russia energy relations more broadly conceived, and in what ways, and why. Some of our findings on the actor landscape on the strategic level and the regional level of northern Europe will be highlighted in the following section. Thereafter, moving onto a higher level of abstraction, it will be indicated what our analysis tells about the possibility of a pan-European energy policy, and especially what it can tell us about the obstacles and hurdles on the way.

Lessons from the Strategic and Regional Level

The first finding is that the well-established approaches of energy diplomacy and energy economics and trade, on the one hand, and geopolitics and energy security, on the other, give us very different but ultimately mutually complementary pictures of the actor landscape and of the actions and interests found therein.

A look at the very active energy diplomacy between the EU and Russia gives us a useful idea of the strong energy interdependence between them. Rational calculations related to energy economics keep EU based and Russian actors interested in each other. Existing and projected investments, and the need for stable energy supplies, especially with a view on Russia's large reserves, will keep EU based actors interested in Russia. And in addition to German energy policy actors, Norwegian companies also increasingly look towards Russia's energy sector. Due to Norway's large remaining reserves, the Norwegians can, however, live with their omission from a shareholder capacity from Russia's attractive Shtokman gas field project.

Russia remains dependent on EU markets in a situation where most pipelines go to the EU, and where the supplies are highly priced compared to supplies elsewhere. Building alternative pipelines is not only costly and time-consuming, but also difficult in the relatively infrastructure and institution free environment in other potential markets in Asia and the Far East. And, Russia is not for some time in a position to supply natural gas on a large scale in the more easily re-directable LNG format. But in the same breath, it is important to note that Russian gas supplies are not likely to grow much to meet any significant increase in EU demand. A slightly higher supply than at present, rising to some 200Bcm from the around 185Bcm in 2006, is likely to be the plateau point for Russian supplies during the next few decades.[1]

A look at geopolitics and the drive for energy security gives us somewhat different insight. From this perspective attention is directed to how EU based and Russian actors try to control the level of their mutual interdependence and the leverage of third parties. For example, Germany and Russia try to minimize the leverage of transit states by means of the Nord Stream gas pipeline project. In this book, Kirsten Westphal portrays the Nord Stream project as a costly investment to increase Germany's energy security. Its prospects of realizing this goal do not look fully convincing, as the supplier country will continue to be the same one – Russia. And, any gains will be obtained at the expense of traditional transit states Ukraine, Belarus and Poland, although some other states and regions such as Sweden may benefit from possible spur lines planned to be built in the second phase of the project.

Geopolitics also elucidates instances where Russian actors have deemed diplomacy insufficient for realizing economic and foreign policy goals. To get an idea of the extent to which energy resources are viewed through the lens of the notoriously sloppy concept of national interests, Stanislav Tkachenko makes the useful observation that the overall interests of Russian foreign policy have assumed more stability during the Putin era than was the case under Yeltsin. Yet, as Tkachenko points out in his chapter, by the end of 2006, no foreign policy document existed in Russia that would cement *how* energy resources are to be used in the service of foreign policy in specific geographical directions and policy areas. The great efforts made to bypass the Baltic transit corridor, as well as Ukraine and Poland as transit states, represent of course important indicators; yet at the same time Tkachenko refers to the underlying economic rationality of reducing transit fees and the uncertainties arising from transit states' political leverage. The geopolitical reading is also present in Tkachenko's and Viatcheslav Morozov's notions of how Russian foreign policy is designed to maintain a sphere of influence within the CIS.

1 Jonathan Stern, 'The New Security Environment for European Gas: Worsening Geopolitics and Increasing Global Competition for LNG', Oxford Institute for Energy Studies, NG 15/2006, pp. 25–7.

The supply cut-offs to Ukraine and Belarus in early 2006 and 2007, respectively, were important indicators of the mixture of geopolitics and economics in Russia's energy relations with the CIS. The failure to secure moderately priced and Russian-controlled transit through Belarus and Ukraine, and to mutually agree on marketizing these energy relationships, prompted Russia to resort to negative sanctions. This kicked off a zero sum game. Tatiana Romanova argues that Ukraine illegally siphoned gas intended for European markets during its cut-off in the 'gas war' with Russia/Gazprom. The consequences are put into critical scrutiny by Viatcheslav Morozov, who suggests that both parties lost in the resulting deal, which created a fully non-transparent regime regardless of Russia's assurances of intending to marketize the relationship. In the end, Russia was on dubious grounds framed as the guilty party in the international media, owing to the widespread perceptions of it simply exercising a traditional geopolitical approach towards its neighbour.

Such perceptions made many Russia-dependent EU member countries threatened.[2] As for the Baltic states, oil supplies to the Ventspils terminal in Latvia were cut off already in 2003. Eiki Berg and Stanislav Tkachenko help to highlight the mutually contradictory Baltic and Russian points of view, and agree that Russo–Baltic transit politics on the whole can be read as a prime example of a traditional geopolitical approach from Russia's part – lessening its dependence on Baltic transit by building expensive new ports onto its own coastline in the Leningrad and Murmansk regions. And in the Baltic side, geopolitical fears entertained by governmental actors make Russian capital not always welcome. However, a cautious attitude to Russian investments represents an all-European phenomenon.[3]

As for the Nordic states, Pami Aalto and Nina Tynkkynen show how the Finnish energy debate is regardless of the otherwise well working Finnish–Russian relations coloured by the high dependency on Russian energy. When this dependency is understood as a threat, it is mostly framed in environmental terms. But implicitly, in some instances at the governmental level, it is taken as a more traditional security threat. On the whole, taking into account the very different perceptions of energy security issues that we have uncovered on the EU and Russian sides, it is comprehensible why the agenda of the EU–Russia energy dialogue has been kept mainly technical and limited. A related development that emerges through the geopolitical and energy security lenses is how both parties are looking to diversify their energy trade towards other partners.

The second finding relates to how a look on energy and the environment helps us to uncover the rather different issue of how the Nordic states, and EU bound actors in general, can afford state-of-the-art climate policies alongside their energy security concerns, although obviously a lot still remains to be done in the sustainability

2 Poland proposed an 'energy NATO' in 2006. For most observers, an energy alliance consisting mostly of countries which are dependent on energy imports does not make much sense.

3 Other examples include the reluctance of E.On Ruhrgas to swap assets with its close ally Gazprom (see Chapter 5). Gazprom's statement in early 2006 of its intention to acquire the British gas company Centrica also became an issue for British governmental policy, although outright blocking of Gazprom's possible bid was ruled out.

sphere within the EU. A further point is how a limited focus on economic rationality cannot fully grasp why attractive energy extraction projects are strongly conditioned by environmental considerations in Norway. It is no less important to note how sustainability issues raised by the Nordics interfere with geopolitically and economically motivated energy projects in the Baltic and Barents Sea areas (Table 9.1).

In addition to these well-established perspectives, our aim was to open up new theoretical approaches to energy policy research and to the study of EU–Russia energy relations. These new approaches offer both wider and nuance richer perspectives in relation to the more commonplace research agendas. The political sociology of energy approach highlights how strong energy interdependence does not automatically translate into corresponding political integration due to incompatibly framed geopolitical models and identities between the EU and Russia. The prospects of building a wider Europe with the EU and Russia as its constituent parts, as well as reforming, modernizing and transforming Russia remain very imperfectly realized. A focus on bureaucratic politics, for its part, shows why energy markets and the energy sector in individual EU member states like Germany remain sub-optimal. The Baltic examples further indicate how there is no guarantee of state involvement necessarily bringing any better results in transit companies' activities than a laissez-faire approach. Finally, a focus on the regional politics of energy helps to dissolve the all too frequent image of Russia as one monolithic energy giant in the making even in conditions of increasing centralization and state supervision in Russia's energy sector (Table 9.2).

Alongside this picture of the actor landscape and of the actions and interests found therein, more generic problems and hurdles for a pan-European energy policy can be pointed out when we move onto a higher level of abstraction. These problems interfere and intersect with future possibilities and can be summarized through the concepts of sovereignty and geopolitics, bilateralism, asymmetry, and the principles applied in energy policy. A wider regional focus than covered in this book could of course yield further issues. But we have strong grounds to expect that the issues we will next highlight represent fairly well generalizable phenomena.

Table 9.1 EU–Russia energy dialogue at the strategic and regional levels (I)

Approach	Premises		
	Actors (who?)	Agency (what?)	Interests (why?)
Energy diplomacy	EU Commission, EU member and associated states, Russia, transit states, and their leaders; leaders of large companies	EU–Russia, EU–Norway, Norway–Russia summits; bilateral German–Russian deals; multilateral regional, ECT, IEA, WTO talks	Strong interdependence between EU and Russia; and between German state and companies, and Russia; Norwegian and Russian companies looking for economic gains; joint EU–Russian interest in a limited/technical common agenda
Geopolitics and energy security	EU Commission, EU member states, Russia, transit states, and their leaders; leaders of state-bound companies	Security of supplies (EU); security of markets (Russia, Norway); cut-off to Ventspils; cut-offs to Ukraine, Belarus, and consequently, reduced supplies to EU; Nord Stream and Murmansk pipelines; new ports	Russian actors pressing transit states for assets, and building new ports in Murmansk and Leningrad region; Russia-dependent Finland, Sweden, Baltic states and Poland looking for supply guarantees and threats; Efforts to diversify EU supplies beyond Russia, and Russia's efforts to diversify markets beyond EU
Energy economics and trade	EU Commission, EU member and associated states, Russia; companies; EBRD, large banks, WTO	In addition to existing means, energy extraction, transport and finance vis-à-vis new fields in northern Norway, northwest Russia and eastern Siberia	EU-based actors' need to access Russia's large reserves when their own are becoming scarce; Russia's reduced but yet real dependence on foreign investment; Norwegian companies' internationalization from a secure domestic position
Energy and the environment	EU Commission, EU member and associated states, Russia, transit states, companies, environmental organizations/NGOs	Kyoto protocol; zero/low emission technologies; Norway's integrated management plan for the Barents Sea; Kirkenes declaration; nuclear safety in northwest Russia	EU and Nordic example in global climate and sustainability policies; threat perceptions of increased oil transports through fragile Barents and Baltic Seas; criticism of the geopolitical character of the Nord Stream gas pipeline vis-à-vis environmental side-effects; questions of sustainability of Russian electricity

Table 9.2 EU–Russia energy dialogue at the strategic and regional levels (II)

Approach	Premises		
	Actors (who?)	**Agency (what?)**	**Interests (why?)**
Political sociology of energy	EU Commission, EU member and associated states, Russia, transit states, and their leaders; companies, organizations/NGOs	Uncertainty, complexity and polarization in EU energy policy, leading to a need for story-telling; rise of identity politics in European energy policy; identity clash between EU and Russia	Balance between securing EU's and Russia's material-economic basis through energy trade, and the more demanding political project of building a wider Europe with EU and Russia as its constituent powers; imperfect utilization of the energy dialogue for Russia's modernization
Bureaucratic politics of energy	EU Commission, EU member and associated states, Russia, transit states, and companies, and bureaucratic actors within them	Bureaucratic struggle within EU institutions; links between German and Baltic governments and companies; bargaining among state institutions in Russia	EU member states' interest in delimiting the Commission's role; state-led building of large energy companies in Norway and Germany, mutually contradictory interests in the Baltic states vis-à-vis transit; the state's and presidential administration's strengthening grip of Russia's energy sector
Regional politics of energy	Russian state, regional and federal institutions, lobbying groups	Centre-region struggles; interregional struggles among old and new energy provinces in Russia	Balance between state-building interests and regional development interests

Sovereignty and Geopolitics

The first problem comes up in all chapters to this book regardless of the theoretical approach employed, and pertains to *sovereignty and geopolitics*. This is by no means entirely unexpected. As an abstract principle, sovereignty is very familiar to those who have studied the self-critical debates within the discipline of international relations during the past one or two decades. It also continues to moderate the praxis of international politics and impacts on political economy.

Concerns related to sovereignty and traditional geopolitics instruct states to opt for national or bilateral solutions in instances where pure economic logic would suggest otherwise. This desire the keep the control over interaction in national hands and not to let too many voices into problem-solving is aggravated by the pattern by which these 'solutions' may not even be optimal for the states concerned, plus that they frequently create problems for other actors, resulting in a problem of the commons.

As for the EU party, sovereignty emerges in how energy supplies and energy mix issues still belong to national hands. An easy and popular prediction proceeds from our opening remarks to this book of how EU member states increasingly take energy as associated with 'high politics' questions that traditionally are linked to national security. This new tendency can further prevent the EU Commission from assuming more power in this sphere and reducing the number of actors operating in the conduct of the EU's external energy relations. But it may not be so simple. The history of EU integration can also, in Alan Milward's terms, be interpreted as a 'European rescue of the nation-state',[4] where many members perceiving an issue crucial for their sovereignty and survival will assign powers to the Union to act when they think their own weight does not suffice to serve best their interests. This may provide the necessary stimulus for deepening energy sector integration. A recent example of the functioning of the same mechanism is how Europe's inability to act on its own backyard in Kosovo in 1999 served to push security co-operation further within the Union,[5] with a debate on common European defence starting shortly afterwards and a rapid reaction force created.

As for the EU–Russia energy dialogue, the question on many European policy-makers' lips is whether Russia's sovereignty concerns and geopolitical interests could one day dictate cut-offs that would be directly and openly intended to the EU party despite three decades of no such interruptions. Pondering on this question represents a very basic precondition to the possibility of pan-European energy solutions. Assuming EU member states' willingness and ability to pay the very likely rising gas prices in the absence of sufficient alternative supplies, answers to the

4 Alan Milward, with the assistance of G. Brennan and F. Romero, *The European Rescue of the Nation-State* (London: Routledge, 2003).

5 Paul Cornish and Geoffrey Edwards, 'Beyond the EU/NATO Dichotomy? the Beginnings of a European Strategic Culture', *International Affairs*, 77/3 (2001): 587–603.

question of cut-offs hinge on the wider foreign policy and foreign economic policy goals adopted in Russia.[6]

Tkachenko's analysis, among many others in this book, suggests that the mutual interdependence is likely to keep Russia pragmatic in the EU direction, striving mainly at economic integration and asset acquisitions rather than forceful domination as may be the case in the CIS. Putting any alarmism to rest, it remains that the EU has to accept sovereignty and geopolitical concerns as a major and possibly strengthening factor in its energy relations with Russia. Instead of dealing with a diverse group of actors ranging from Russian regions and companies to ministries as seemed the way ahead in the 1990s, the EU will have to deal with the Russian state. However, within Russia itself, competition will continue, especially between the oil and gas sector actors, and new energy provinces and transit regions. This all suggests that Russia's geopolitical interests will primarily hit the transit states with which the EU is developing closer relations with its new neighbourhood policy launched in 2003,[7] like Ukraine and the Caucasus, but also 'black holes' like Belarus.

As for our introductory thesis of northern Europe representing a crucial case study with regard to the functioning of the regional level vis-à-vis the grander level of the EU–Russia energy dialogue, the implications are now clear. Sovereignty concerns are making energy securitized into a 'high politics' issue. Consequently, it is becoming too 'big' to be dealt with successfully at the regional level alone. Our finding of the northern European energy agenda being mainly set at the strategic level and driven by large states and corporations suggests precisely this. Energy projects are costly, requiring investments from large corporations and involvement of international financial institutions, and in many cases attract interest from extra-regional state actors. This means that most likely, there is no stand-alone regional energy agenda in sight for northern Europe or any other sub-region that would differ notably from the bigger picture. But implementation of course remains something to be done very visibly at the regional level. Regional organizations may assume important functions in supporting activities, as indicated by the role of the BASREC in Romanova's chapter, and the planned ND partnership in logistics and transport, and the possibility of an energy partnership as mentioned by Aalto and Tynkkynen.

Regional level actors located between the EU and Russia should first look at the all-European level and the Russian direction in order to determine the interests of larger scale actors and consequently discern which options and strategies are available for themselves. This does not mean that they might not be able to propose new policy openings, transport routes, pipelines, storage spaces, energy regimes, and the like, but that the realization of their proposals is unlikely within the regional

6 Some authors like Debra Johnson find competing policy models in Russia's use of energy in its foreign policy, with the sovereignty-bound geopolitical model dominating in pipeline questions. However, like we do as well, Johnson suggests that the notion of interdependence between Russia and the EU grasps well the overall dominating model in their energy relations; see Debra Johnson, 'EU–Russia Energy Links: A Marriage of Convenience?', *Government and Opposition*, 40/2 (2005): 276–7.

7 Pami Aalto, 'The EU, Russia and the Problem of Community', in Hartmut Mayer and Henri Vogt (eds), *A Responsible Europe? Ethical Foundations of EU External Affairs* (Basingstoke: Palgrave, 2006), pp. 109–12.

confines. A further point of note is that other regions than the north European one are not likely to possess that many assets as does northern Europe, which we deemed at the outset a promising setting for regional energy co-operation.

Asymmetry

The sovereignty and geopolitics issue prompts the second problem for a pan-European energy policy on what types of sovereigns there are. Sovereignty assigns a formal status to territorial entities whereby they are admitted to international clubs such as northern Europe's regional organizations, and the EU, as equals. In practice this is a fallacy due to the persisting *asymmetry*, resulting from the unequal division of power resources, including energy in the present agenda, even in contexts where borders are disappearing like the EU.[8] In the northern European regional context that we have used as an example, this translates into less empowered voices of the Finns and Balts who have a vested interest in regional energy arrangements in the sphere of supply and transit from Russia. But as implied, they will have to accept that a large part of the agenda of EU–Russia energy relations is set at the strategic level, coupled with bilateral projects by major players like Germany and Russia, and possibly in the future, energy major power Norway with Russia. The Finnish perceptions of the EU Commission listening the regional level too little, and talking too little, or in the wrong way to Russia, are good examples of what may be ahead even more, when energy questions become hot topics in big tables. Sweden, eying more energy supplies from Russia, might find itself in the same situation, as could other small EU members.[9] The countries of the Caucasus may surprisingly be somewhat better situated as they offer an alternative transit route for energy resources from the greater Middle Eastern area and Central Asia whilst also bypassing Russia.

Bilateralism

The third problem arises from the noted *bilateral solutions*. This is an issue with which the EU struggles more generally as well in its relations with Russia. But as claimed in a recent volume, somewhat surprisingly, the bilateral track may in fact be a useful additional channel, especially when the strategic partnership is temporarily strained, and in situations where the enlarged EU has a specific difficulty to agree on its Russian policy, with Poland and the Baltic member states in particular more sceptical of Russia than many old members. Furthermore, bilateral relations do not necessarily exclude concomitant action by the same state at the EU level.[10] But the Nord Stream gas pipeline case suggests that this logic may not work in energy policy,

---

8 See also Jyrki Käkönen, 'Asymmetry and Interaction: Borders in the International System', in Heikki Eskelinen, Ilkka Liikanen and Jukka Oksa (eds), *Curtains of Iron and Gold: Reconstructing Borders and Scales of Interaction* (Aldershot: Ashgate, 1999).

9 See Robert. L. Larsson, *Sweden and the NEGP: A Pilot Study of the North European Gas Pipeline and Sweden's Dependence on Russian Energy* (Stockholm: Defence Research Agency, June 2006).

10 Jackie Gower, 'Conclusion: EU Foreign Policy as a Two-Level Game', in Hanna Smith (ed.), *The Two-Level Game: Russia's Relations with Great Britain, Finland and the European Union* (Helsinki: Aleksanteri papers 2/2006).

which, as shown throughout the book, tends to spill over from one policy sector to another. Not only is economics mixed with perceptions of history and identity, as in Russo–Baltic relations, and to an extent in Norway and Finland, as well as with politics, and also, traditional geopolitics as in the case of Russia, but the environment also frequently enters the picture. Neither do bilaterally conducted energy projects stay within their intended confines, but tend to draw in other interested parties perceiving their energy security to be at stake. The proposed electricity cable from Russia to Sweden sparked an openly politicized debate in the suggested transit country Finland, and alongside the Nord Stream pipeline represents a further such example. But not always does the energy policy spill-over work in the same way. To take one example, Latvia was unable to successfully internationalize its own transit problems in Ventspils. Although we suggested in the introduction that energy problems should be seen as shared problems, this perception does not always arise in political praxis. Here we return to the noted asymmetries.

Commitment to Principles

Sketching the fourth problem should start with our finding of the prevalence of an *imperfect commitment to joint principles among EU member states*. The EU exercises considerable influence in defining the European energy agenda, but the three principles that the EU Commission promotes are not uniformly applied in the member states. Along the markets and competition principle, the German market is open but remains oligopolist and state intervened. In the rest of the northern EU member states, only the Nordic electricity market satisfies well the markets and competition principle. Along the security of supplies principle, Denmark is the only EU member state not expressing the fears of future supplies fuelled by relative energy poverty. In Germany, Russian supplies are viewed with relatively little concern to diversify sources. But in Finland, nuclear power is enjoying a new lease of life in order to reduce the Russian dependency, and doubts are raised even about an electricity cable from Russia which would provide more competition and lower prices. Sweden is drifting into a greater Russian dependence. In the Baltic states the picture is mixed both country-wise and along energy sources. Along the sustainability principle, Germany and the Nordics manifest high awareness of the issue and a reasonable readiness to act, albeit doubts are raised about the sustainability of Finland's imported electricity. In the Baltic states, the principle is weakly internalized (Table 9.3).[11]

On the whole, these northern European cases illustrate the wider heterogeneity within the EU with regard to commitment to the principles of common energy policy. And looking further from the Union's point of view, the imperfect commitment to the principles promoted by the EU is cast in a more serious light when we move to summarizing the picture in directions where the EU's most important supplies are.

11 Source: Chapters 1–8; European Commission, 'The Commission to Act Over EU Energy Markets', memo/06/481, 12 December 2006, Brussels, <http://europa.eu/rapid/pressReleasesAction.do?reference=MEMO/06481&format=HTML&aged=1&language=EN&guiLanguage=en> (accessed 23 March 2007).

Table 9.3 Principles of EU energy policy and member states' commitment

Actor	Principle of EU energy policy		
	Markets and competition	Security of supplies	Sustainability
EU Commission	Key idea of EU integration; EU's status as a 'regulatory state' in charge of fair and functioning markets; pro-ECT; anti-monopolism; interconnection of grids and third-party access; marketized prices	Energy poverty; concerns of energy mix; diversification of supplies new concern; new crisis management rules and solidarity, storage systems; savings, efficiency; alternative energy sources emerging issue	Included since 1991; fight against climate change, pro-Kyoto protocol and reduction of CO_2 emissions; energy savings, efficiency; new technologies for CO_2 capture
Germany	Pro-ECT; liberalized gas and electricity markets by 1997–8, but oligopolism and state involvement; new bilateralism; companies in Russia	Energy poverty; pro-energy efficiency, renewables; Nord Stream pipeline bypassing transit states with imperfect added value and high cost	EU compatible position, or even at the forefront of EU sustainability policies
Finland	Electricity market part of Nordpool and linked to the Baltic states; Fingrid fearing new competitors and reduced prices; gas market not competitive; Fortum active in Russia	Concerns for the 40% dependence on Russia, framed in environmental and traditional security terms; nuclear reactors as solution; logistics, transport and possibly energy to ND	Concerns of increasing oil transports through Baltic Sea and nuclear safety in Russian power plants; energy–environment link both at governmental and NGO level
Sweden	Electricity market part of Nordpool; insufficient unbundling of distribution and transmission, and network access; Vattenfall in Germany and Poland	Energy policy stalemate after nuclear phase-out; increasing dependence on Russian oil, possibly electricity and gas	High share of domestic sustainable energy; concerns of environmental impact of the Nord Stream pipeline; increasing imports often non-sustainable
Denmark	Electricity market part of Nordpool; functioning energy market; gas exported to Sweden but no major energy companies	No concerns in the mid to long-term; security of demand in Sweden	High share of domestic sustainable energy; high environmental taxes; sustainable technology industry
Estonia	Regulated prices, lack of unbundling, access to networks; laissez-faire approach to energy transit	Self-sufficient in electricity; domestic oil shale; 100% dependence on Russian gas	Weakly internalized; yet concerns of oil accidents; NGOs weak on energy
Latvia	Regulated prices but unbundling completed; state strategy on energy transit, but mixed record	40% dependent on imported electricity; dependent on oil imports and Russian gas	Weakly internalized; transit through Latvia accepted even by environmentalists
Lithuania	Imperfect unbundling, price marketization, network access; moderate state involvement in transit	Plans for a (Baltic) nuclear plant after Ignalina; 100% dependent on Russian gas	Weakly internalized; yet concerns of oil drilling in Kaliningrad's sea shelf

That is, there is *even more dissonance vis-à-vis the principles of EU energy policy in the wider European area*, where the EU first attempted to spread them through the ECT, and now through the EU–Russia and EU–Norway energy dialogues.
We find the Putin regime and Russia's gas sector at odds with the EU's markets and competition agenda, coupled with a strengthening state involvement in Russia's still largely privatized oil sector. Alas, EEA member Norway performs very well here for a major energy producer. The security of supplies principle does not make sense in the same way for Russian and Norwegian actors as it does for actors in the EU area. At issue for them is security of markets and demand, although the Norwegians were found to mention the security of supplies principle when speaking to the Union's representatives. Some evidence of an instrumental and case-by-case adoption of the sustainability principle is found on the part of the Russian state, Gazprom and oil companies. But Norway is a pioneer in this field, again somewhat surprisingly for a major energy producer (Table 9.4).

On the basis of this summary analysis of the generic problems within EU–Russia energy dialogue and its north European level, it is clear that apart from the underlying strong interdependence and the intense energy trade that is going on a day-to-day basis in the wider European area, there is a lot to do before we have a pan-European energy policy in place. We have pan-European energy trade, but we do not have pan-European energy policy with the characteristic assumptions for the term 'policy' – as connoting a degree of direction, purpose and stability of energy politics, and pulling along governmental and foreign policies within the wider European area. In this situation, it is understandable that projects like the ECT still await for ratification, and do not look like the way ahead in their existing form, if the present multitude of policies in the wider European area stays as incompatible as we have indicated.

The greatest discontinuities are between the EU and Russia, who paradoxically at the same time represent the highest potential for a mutually compatible major energy relationship. But regardless of their mutual discontinuities, they manage to stage a much more promising setting than is found in most regions of the world, for example in the Far East. Over there Russia is a very potential supplier for the energy poor Japan and South Korea, and China with its phenomenally growing energy needs, thanks to Russia's riches from the eastern Siberian fields and the Sakhalin Island. But in this regional context, the practical realization of the potentially compatible energy relationship is difficult in the absence of any regional co-operation frameworks, institutions, infrastructure, multilateralism for reducing risks in expensive and technically demanding projects,[12] and the like, all of which are found in the wider European area, albeit most of them in a yet imperfect form.

12 See Aleksanteri Institute Eurasia Energy Group, 'The Energy Dynamic on the Borders of the EU: Internal Dynamics of the Russian Energy Sector', policy memo, <www.helsinki.fi/aleksanteri/energy>.

Table 9.4 Principles of EU energy policy plus Russian and Norwegian actors

	Principle of EU energy policy		
Actor	**Markets and competition**	**Security of supplies**	**Sustainability**
The Putin regime	State ownership and/or overseeing, including pipeline control; no to ECT; restricting access to resources, refusing to grids; bypassing transit countries	Security of demand, prices and contracts in Europe	EU-induced, instrumental joining of Kyoto protocol; energy savings and efficiency piloted in the EU–Russia energy dialogue and ND
Russia's gas sector	State protection in upstream and domestic markets, yet in downstream, striving for strengthened market access and acquisitions on the basis of EU internal market rules	Security of demand, prices and contracts in Europe and prospectively, Asia	Some evidence of commitment to sustainability in particular in Norwegian–Russian energy dialogue
Russia's oil sector	State's efforts to make its own companies competitive/ privileged vis-à-vis globally and locally operating private Russian companies; toughening conditions for foreign companies	Security of demand, prices and contracts in Europe and globally	Some evidence of commitment to sustainability in building the new Baltic Sea ports and in Timan Pechora especially in joint ventures
Regions in North-western Federal District	Intense competition between regions for state and company projects in the energy sector; state support increasingly important for energy revenues of regions	Security of demand, prices and contracts in Europe (and beyond)	Especially for poorer regions, energy revenues override local environmental cost
Norway	Electricity market part of Nordpool; state-guided opening of Norway's continental shelf for foreign companies; internationalization of Norwegian energy companies	Security of demand, prices and contracts in Europe (in the case of gas) and globally (in the case of oil; LNG prospectively)	Strong commitment to sustainability; internal CO_2 tax before Kyoto; zero emissions technology in Barents energy extraction and in gas fired power plants

Source: Chapters 1–8.

The obvious question that arises from this sketchy global level comparison is: taken that conditions for the EU–Russia energy relationship nevertheless do not look that bad as they might, what can be done to make it succeed in the eyes of both parties? This is a key question with regard to envisaging building blocks for a pan-European energy policy, and one that a book with a policy agenda like ours has to try to answer. We propose two policy options. They may look simple and obvious by now, but for the sake of setting some signposts for future debate they are worth a discussion.

Towards a Reinforced Dialogue?

The first option is to keep the present structures, aims and ambition level of the EU–Russia energy dialogue for the most part as they are. This option reproduces the conventional policy analytic approach of seeking compromises and incremental progress by approximating the parties' positions in a tit-for-that type of game.[13] The way in which the EU and Russia reached the compromise on the Kyoto protocol ratification issue provides a successful example of this option as applied into the sphere of energy policy. As described in this book by Aalto and Romanova, after protracted negotiation rounds, the EU got Russia's ratification of the Kyoto protocol, and a promise to gradually abolish differential pricing for domestic and foreign energy buyers, whilst Russia got the EU's support for its all-important WTO membership bid.

But considering how far the EU's and Russia's positions were found to be from each other in the discussion above, in particular along the markets and competition principle, and how unlikely their rapid approximation was deemed, it is difficult to see how choosing this option could lead to a joint agenda eventually kick-starting a pan-European energy policy. It is particularly difficult to see how this option could draw along the needed regional and other smaller scale actors, including transit states. Instead, it is more likely that we would see a familiar pattern of small conflicts and compromises allowing the energy trade negotiations to continue in the present piecemeal fashion between the EU Commission and Russia. More fundamental questions would remain unanswered, or even purposefully ignored as they are now. The long-term foundation for the energy relationship could also become in jeopardy. Slow progress could lead to the continued absence of the required large scale investments from the European side into Russia's energy sector. Mutual economic integration would be in danger of slowing down, and Russia's energy sector in danger of developing sub-optimally. This all combined could make Russia unable to meet the huge expectations of its growing clientele. Finally, in this scenario, it is somewhat difficult to foresee the EU maintaining its present absolutely privileged customer position in the very long term.

The second option is to boldly expand the actor landscape of the EU–Russia energy dialogue to cover better the interests of member states, including also the smaller ones among them, and in this way to involve them better. Aside the currently

13 See Emery Roe, *Narrative Policy Analysis: Theory and Practice* (Durham: Duke University Press, 1994), p. 18.

included large business actors, the Union could additionally consider ways of involving smaller enterprises, and also, non-governmental organizations with an interest in the dialogue. These measures would intensify the linkage between the strategic and regional levels,[14] as the primary concerns of these actors are usually found on the regional level. Especially this remark applies to those actors located in important transport interfaces or implementation areas of the energy dialogue. Another implication would be the expansion of the scope of policy sectors, away from the technicalities and economic issues, and towards the wider political, environmental and identity political questions of concern to the smaller actors, and offering new departures for the EU and Russia as suggested by Aalto and Morozov in this book.

The sectoral expansion would naturally make the agreed agenda much more complex instead of facilitating compromises in the short term. But in policy questions that are already *de facto* full of uncertainty and complexity, and polarization among and within the parties, this may be a strategy worth thinking of seriously.[15] As we trust to have shown convincingly, sectoral expansion towards the wider ramifications of the energy dialogue provides new opportunities in thinking of political incentives like finding a more integral place for Russia in Europe, and for Russia to benefit from the dialogue outside the narrow energy sales sector by technology transfers and increased interaction in ever more sectors to modernize its economy, industries, and ultimately, its social organization. All of these are outspoken goals of the present Russian regime, although here it must be noted that a non-dialogic or unidirectional Europeanization is definitely not what we intend here. A related aspect is that these wider cross-sectoral linkages *are* already involved in EU–Russia energy relations, in particular in the implementation side that takes place at the regional level. In this sense, sectoral expansion would make a now implicit factor more explicit. But the sectoral expansion would need to be mutual, in the sense of the EU side also learning to listen to Russia's issue linkages and being ready learn from Russia.

* * * *

Regardless of which policy option is chosen for the further development of the EU–Russia energy relations that must play a key part in any pan-European energy agenda, we maintain that the choice should be done deliberately. By our short discussion above we hope to have facilitated the choice, although the responsibility for practical action rests with the political actors themselves. As for the responsibilities of the academic community, as a very final remark, it should be clear by now that making sense of EU–Russia energy relations, and of energy policy in a wider sense, necessitates keeping the actor landscape and sectoral view wide and inclusive in terms of how the whole research agenda is defined.

---

14 To an extent this is taking place at the more general plane in the EU side, where regional scope policies like the ND and the neighbourhood policy are becoming more closely linked to EU foreign policy activities; see Aalto, 'The EU, Russia and the Problem of Community'.

15 Roe, *Narrative Policy Analysis*, pp. 1–19, especially 18–19.

Bibliography

Aalto, Pami, 'The EU, Russia and the Problem of Community', in Hartmut Mayer and Henri Vogt (eds), *A Responsible Europe? Ethical Foundations of EU External Affairs* (Basingstoke: Palgrave, 2006).

Aalto, Pami, *European Union and the Making of a Wider Northern Europe* (London: Routledge, 2006).

Adomeit, Hannes, Bastian, Katrin and Götz, Roland, 'Deutsche Russlandpolitik unter Druck', *SWP Aktuell*, 56 (2004).

Aleksanteri Institute Eurasia Energy Group, 'The Energy Dynamic on the Borders of the EU: Internal Dynamics of the Russian Energy Sector', policy memo <www.helsinki.fi/aleksanteri/energy>.

Al-Kasim, Farouk, *Managing Petroleum Resources: The 'Norwegian Model' in a Broad Perspective* (Oxford: Oxford Institute for Energy Studies, 2006).

Allison, Graham T., *Essence of Decision: Explaining the Cuban Missile Crisis* (Boston: Little, Brown and Company, 1971).

Arbatov, Alexander, Feygin, Vladimir and Smirnov, Victor, 'Unrelenting Oil Addiction', *Russia in Global Affairs*, 3/2 (2005), 142–55.

Baev, Pavel K., 'Chairing the G8: Russian Energy and Great Power Aspirations', *PONARS Policy Memo*, no. 382 (2005), <http://www.csis.org/media/csis/pubs/pm_0382.pdf>.

Balmaceda, Margarita M., 'EU Energy Policy and Future European Energy Markets: Consequences for the Central and East European States', Arbeitspapiere, MZES 42/2002.

Barnes, Joe, Jaffe, Amy Myers and Morse, Edward L., 'Geopolitics of Russian Supply and US Foreign Policy', in *The Energy Dimension in Russian Global Strategy* (The James A. Baker III Institute for Public Policy of Rice University, 2004) <http://www.rice.edu/energy/publications/russianglobalstrategy.html>.

Barysch, Katinka, 'EU–Russia Economic Relations', in Oksana Antonenko and Kathryn Pinnick (eds), *Russia and the European Union* (London: International Institute for Strategic Studies, 2005).

Bastian, Katrin and Götz, Roland, 'Unter Freunden? Die deutsch-russische Interessenallianz', *Blätter für deutsche und internationale Politik*, 5 (2005), 583–92.

Berg, Eiki, 'Some Unintended Consequences of Geopolitical Reasoning in Post-Soviet Estonia: Texts and Policy Streams, Maps, and Cartoons', *Geopolitics*, 8/1 (2003), 101–20.

Brodin, Alf, 'Breaking Loose? Russian Reasons to End Its Dependence on Foreign Ports', in Lars Hedegaard and Bjarne Lindström (eds), *The NEBI Yearbook*

2001/2002: North European and Baltic Sea Integration (Berlin: Springer Verlag and Nordegio, 2003).

Brodin, Alf, *Ports in Transition in Countries in Transition* (Gothenburg: Centre for European Research, Gothenburg University, Report 8/2000).

Brown, Archie, 'Vladimir Putin's Leadership in Comparative Perspective', in Cameron Ross (ed.), *Russian Politics under Putin* (Manchester: Manchester University Press, 2004).

Browning, Christopher S., 'The Internal/External Security Paradox and the Reconstruction of Boundaries in the Baltic: The Case of Kaliningrad', *Alternatives,* 28/5 (2003), 545–81.

Browning, Christopher S., 'Westphalian, Imperial, Neomedieval: The Geopolitics of Europe and the Role of the North', in Christopher S. Browning (ed.), *Remaking Europe in the Margins: Northern Europe after the Enlargements* (Aldershot: Ashgate, 2005).

Cecchini, Paolo, Jones, Erik and Lorentzen, Jochen, 'Europe and the Concept of Enlargement', *Survival*, 43/1 (2001), 155–65.

Chatelus, Michel, 'Oil Security and Relations between the Gulf Countries and the EU', in Michel Chatelus, Giacomo Luciani and Eberhard Rhein (eds), *EU-GCC Co-operation in the Field of Energy*, EUI Policy Paper, 2/3.

Chow, Edward C., 'Rossijskie truboprovody: Nazad v buduschee?' *Pro et Contra*, 8/3 (2004), 164–73.

Christian Meier, 'Deutsch-Russische Wirtschaftsbeziehungen unter Putin, Praxis-Probleme-Perspektiven', *SWP Studie*, 41 (November 2004).

Clingendael International Energy Programme, 'Study on Energy Supply Security and Geopolitics', final report prepared for DG TREN, The Hague, January 2004.

Cordesman, Anthony H. and Al-Rodhan, Khalid R., *The Changing Risks in Global Oil Supply and Demand: Crisis or Evolving Solutions?* (Washington: Centre for Strategic and International Studies, 2005) <http://www.csis.org/media/csis/pubs/050930_globaloilrisks.pdf>.

Cornish, Paul and Edwards, Geoffrey, 'Beyond the EU/NATO Dichotomy? The Beginnings of a European Strategic Culture', *International Affairs*, 77/3 (2001): 587–603.

Correlje, Aad and van der Linde, Coby, 'Energy Supply and Geopolitics: A European Perspective', *Energy Policy*, 34 (March 2006), 532–43.

Dinan, Desmond, *Europe Recast: A History of the European Union* (Basingstoke: Macmillan, 2004).

Dryzek, John, *The Politics of the Earth: Environmental Discourses* (Oxford: Oxford University Press, 1997).

Eckstein, Harry, 'Case Study and Theory in Political Science', in Fred I. Greenstein and Nick W. Polsby (eds), *Handbook of Political Science, Vol 7: Strategies of Enquiry* (Reading, MA: Addison–Wesley, 1975).

Eesti transiidikoridori konkurentsivõime ja selle tõstmise võimalused (Tallinn: Strateegiliste Algatuste Keskus, 2003/2004).

Eriksson, Johan, 'Security in the Barents Region: Interpretations and Implications of the Norwegian Barents Initiative', *Cooperation and Conflict*, 30/3 (1995), 259–86.

Foley, Gerard and Lönnroth Måns, 'The European Transition from Oil: Mapping the Landscape', in Gordon T. Goodman, Lars A. Kristoferson and Jack M. Hollander (eds), *The European Transition from Oil: Societal Impacts and Constraints on Energy Policy* (London: Academic Press, 1981).

Gillingham, John, *Coal, Steel and the Rebirth of Europe, 1945–1955* (Cambridge: Cambridge University Press, 1991).

Gillingham, John, *European Integration, 1950–2003: Superstate or New Market Economy?* (Cambridge: Cambridge University Press, 2003).

Goodman, Gordon T., 'Preface', in Gordon T. Goodman, Lars. A. Kristoferson and Jack M. Hollander (eds), *The European Transition from Oil: Societal Impacts and Constraints on Energy Policy* (London: Academic Press, 1981).

Gower, Jackie, 'Conclusion: EU Foreign Policy as a Two-Level Game', in Hanna Smith (ed.), *The Two-Level Game: Russia's Relations with Great Britain, Finland and the European Union* (Helsinki: Aleksanteri Papers 2/2006).

Grace, John, *Russian Oil Supplies: Performance and Prospects* (Oxford: OUP for the Oxford Institute for Energy, 2005).

Götz, Roland, 'Rußlands Energiestrategie und die Energieversorgung Europas', *SWP Studie*, 6 (2004).

Götz, Roland, 'Schweigen für Gas? Deutschlands Abhängigkeit von Rußlands Energielieferungen', *SWP Aktuell*, 43 (September 2004).

Götz, Roland, 'Die Ostseegaspipeline – Instrument der Versorgungssicherheit oder politisches Druckmittel?', *RusslandAnalysen*, 72 (2005), 12–15.

Haukkala, Hiski, 'A Problematic "Strategic Partnership"', in Dov Lynch (ed.), *EU-Russia Security Dimensions* (Paris: European Union Institute for Security Studies, 2003).

Heinrich, Andreas, *Globale Einflussfaktoren auf das Unternehmensverhalten* (Münster: LitVerlag, 2004).

Hill, Fiona, *Energy Empire, Oil, Gas, and Russia's Revival* (London: Foreign Policy Centre, 2004).

Hiltunen, Heidi, *Finland and Environmental Problems in Russia and Estonia* (Helsinki: The Finnish Institute of Foreign Affairs, 1994).

Hønneland, Geir, 'Identity Formation in the Barents Euro-Arctic Region', *Cooperation and Conflict*, 33/3 (1998): 277–97.

Hänninen, Saara, and Rytkönen, Jorma, *Oil Transportation and Terminal Development in the Gulf of Finland* (Helsinki: VTT Publications, 2004).

Häyrynen, Nina, 'Changing Strategies of Environmental Co-operation between Finland and Russia', in *Northern Veche: Proceedings of the Second NRF Open Meeting* (Akureyri: Northern Research Forum, 2004).

Isakova, Irina, *Russian Governance in the Twenty-First Century: Geo-strategy, Geopolitics and Governance* (London: Frank Cass, 2005).

Jaffe, Amy Myers and Manning, Robert, 'Russia, Energy and the West', *Survival*, 43/2 (Summer 2001): 133–52.

Joenniemi, Pertti (ed.), *Saint-Petersburg – Russian, European and Beyond* (St. Petersburg: St. Petersburg State University Press, 2001).

Joenniemi, Pertti and Browning, Christopher S. 'Discourses on Centrality and Marginality: The European Neighbourhood Policy and Finnish Options of

Europe-Making', paper presented in the VII World Congress of ICCEES, Berlin, 25–30 July 2005.

Johnson, Debra, 'EU–Russia Energy Links: A Marriage of Convenience?', *Government and Opposition*, 40/2 (2005), 256–77.

Johnson, Debra, 'EU–Russia Energy Links', Debra Johnson and Paul Robinson (eds), *Perspectives on EU–Russia relations* (London: Routledge, 2005).

Johnson, Debra and Robinson, Paul (eds), *Perspectives on EU–Russia Relations* (London: Routledge, 2005).

Käkönen, Jyrki, 'Asymmetry and Interaction: Borders in the International System', in Heikki Eskelinen, Ilkka Liikanen and Jukka Oksa (eds), *Curtains of Iron and Gold: Reconstructing Borders and Scales of Interaction* (Aldershot: Ashgate, 1999).

Khristenko, Viktor, 'Nuzhna li nam integratsia?' *Rossia v globalnoi politike*, 2/1 (2004), 74–87.

Körner, Stefan 'Instrumente der Energiepolitik', in Danyel Reiche (ed.), *Grundlagen der Energiepolitik* (Frankfurt am Main: Peter Lang, 2005).

Kostianoy, A., Lebedev, S., Litovchenko, K., Stanichny, S. and Pichuzhkina, O., 'Oil Spill Monitoring in the Southeastern Baltic Sea', *Environmental Research, Engineering and Management*, 33/3 (2005), 73–9.

Kusznir, Julia and Pleines, Heiko, 'Reichtum für wen? Politische Ökonomie rußländischer Ölfelder', *Osteuropa*, 9–10 (2004), 340–54.

Lane, Thomas, 'Lithuania: Stepping Westward', in Artis Pabriks, Aldis Purs, Thomas Lane and David Smith (eds), *The Baltic States: Estonia, Latvia and Lithuania* (London: Routledge, 2002).

Larsson, Robert L., *Sweden and the NEGP: A Pilot Study of the North European Gas Pipeline and Sweden's Dependence on Russian Energy* (Stockholm: Defence Research Agency, June 2006).

Laurila, Juhani, 'Transit Transport between Russia and the European Union in Light of Geopolitics and Economics', *Emerging Markets Finance and Trade*, 39 (2003), 27–57.

Lausala, Tero, 'The Role of Energy in the Northern Dimension', in Lassi Heininen (ed.), *Northern Borders and Security – Dimensions for Regional Cooperation and Interdependence* (Turku: Turku School of Economics and Business Administration, Business Research and Development Centre, 2002).

Lipgens, Walter, *A History of European Integration, Vol I, 1945–47* (Oxford: Clarendon Press, 1982).

Loskot, Agata 'Security of Russian Gas Supplies to the EU – the Question of Infrastructural Connections', *Policy Briefs*, Center for Eastern Studies, Warsaw (2005).

Mabro, Robert, 'The Peak Oil Theory', *Oxford Energy Comment*, September 2006.

Majone, Giandomenico, 'Regulatory Federalism in the European Community', *Environment and Planning C: Government and Policy*, 10 (1992), 299–316.

Marin, Anaïs, 'Integration without Joining? Neighbourhood Relations at the Finnish-Russian Border', *DIIS Working Paper*, 14, EUBorderConf series, <http://www.diis.dk/graphics/Publications/WP2006/WP2006-14.pdf>.

Matlary, Janne Haaland, 'The Nordic Countries and EU Membership: the Energy Factor', in Lee Miles (ed.), *The European Union and the Nordic Countries* (London: Routledge, 1996).

McFaul, Michael, Petrov, Nikolai and Ryabov, Andrei, *Between Dictatorship and Democracy: Russian Post-Communist Political Reform* (Washington DC: Carnegie Endowment for International Peace, 2004).

McLean, Iain, 'Two Analytical Narratives about the History of the EU', *European Union Politics*, 4/4 (2003), 499–506.

Mikkel, Evald and Pridham, Geoffrey, 'Clinching the "Return to Europe": the Referendums on EU Accession in Estonia and Latvia', *West European Politics*, 27 (2004), 716–48.

Milward, Alan, with the assistance of G. Brennan and F. Romero, *The European Rescue of the Nation-State* (London: Routledge, 2003).

Moravcsik, Andrew, *The Choice for Europe: Social Purpose and State Power from Messina to Maastricht* (Ithaca, NY: Cornell University Press, 1998).

Morozov, Viatcheslav, 'New Borderlines in a United Europe: Democracy, Imperialism and the Copenhagen Criteria', in *Russia's North West and the European Union: a Playground for Innovations* (Nizhny Novgorod: R-US Expert Transit and Danish Institute for International Studies, 2005).

Morozov, Viatcheslav, 'The Discourses of St. Petersburg and the Shaping of a Wider Europe: Territory, Space and Post-Sovereign Politics', *COPRI Working Papers*, 13/2002.

Mouritzen, Hans, 'Security Communities in the Baltic Sea Region: Real and Imagined', *Security Dialogue*, 32/3 (2001), 297–310.

Müller, Friedemann, 'Klimapolitik und Energieversorgungssicherheit: Zwei Seiten derselben Medaille', *SWP-Studie*, 14 (April 2004).

Müller, Werner, 'Energiepolitik für die Zukunft', in Jürgen F. Baur (ed.), *Die Energiewirtschaft im Umbruch, Rechtliche und politische Probleme im neuen Wettbewerb* (Baden-Baden: Nomos, 2000).

Myrjord, Anne, 'Governance beyond the Union: EU boundaries in the Barents Euro-Arctic Region', *European Foreign Affairs Review*, 8 (2003), 239–57.

Neumann, Iver B., 'A Region-Building Approach to Northern Europe', *Review of International Studies*, 20 (1994), 53–74.

Neumann, Iver, 'Ringmar on Identity and War', *Cooperation and Conflict*, 32 (1997), 309–30.

Nuscheler, Ulrike, 'Energiepolitik', in Werner Weidenfeld and Wolfgang Wessels (eds), *Jahrbuch für Europäische Integration 2002/2003* (Berlin: Nomos, 2003).

Ojala, Lauri, Naula, Tapio and Hoffmann, Torsten, 'Trade and Transport Facilitation Audit of the Baltic States: On a Fast Track to Economic Development' (Washington, DC: The World Bank, 2005).

Olcott, Martha Brill, 'Vladimir Putin i neftyanaya politika Rossii', *Carnegie Center Moscow Working Papers*, 1 (2005).

Palonen, Kari, 'Introduction: Reading the Political: From Policy and Polity to Politicking and Politicization', in Kari Palonen and Tuija Parvikko (eds), *Reading the Political: Exploring the Margins of Politics* (Helsinki: Finnish Political Science Association, 1993).

Pelto, Elina, 'Environmental Risk of the Increasing Oil Transportation in the Gulf of Finland', in Kari Liuhto (ed.), *Growing Russian Oil Shipments in the Baltic Sea: Strategic Decision or Environmental Risk?* (Lappeenranta: Lappeenranta University of Technology Publication, 2003).

Peters, Susanne, 'Courting Future Resource Conflict: The Shortcomings of Western Response Strategies to New Energy Vulnerabilities', *Energy Exploration & Exploitation*, 23/1 (2003), 29–60.

Putin, Vladimir, 'Mineralno-syrievye resursy v strategii razvitija rossijskoj ekonomiki', *Notes of the Mining Institute*, 144/1 (St. Petersburg: Saint-Petersburg Mining Institute, Technical University, January 1999), 3–9.

Reiche, Danyel, 'Governance und Energiepolitik', in Danyel Reiche (ed.), *Grundlagen der Energiepolitik* (Frankfurt am Main: Peter Lang, 2005).

Remington, Thomas F., *Politics in Russia* (New York: Longman, 1999).

Roe, Emery, *Narrative Policy Analysis: Theory and Practice* (Durham: Duke University Press, 1994).

Romanova, Tatiana, *Stanovlenie Evropeiskogo soyuza kak mezhdunarodnogo aktora: Na primere investitsionnoi deyatelnosti v energeticheskom sotrudnichestve s Rossiei 1994–2001gg* (Sankt-Peterburg: Izdatelstvo Sankt-Peterburgskogo universiteta, 2003).

Romanova, Tatiana, 'EU-Russian Energy Dialogue and Northern Dimension' in L. Hedegaard and B. Lindström (eds) *The NEBI Yearbook 2004: North European and Baltic Sea Integration* (Berlin: Springer, 2005).

Rutland, Peter, 'Putin and Oilgarchs', in Dale R. Herspring (ed.), *Putin's Russia: Past Imperfect, Future Uncertain* (Lanham: Rowman & Littlefield Publishers, 2005).

Rybakov, F.F., *Ekonomika Sankt-Peterburga: proshloye, nastoyaschee, budischee* (St. Petersburg: Gidrometeoizdat, 2000).

Rytövuori-Apunen, Helena, 'Forget "Post-Positivist" IR!: The Legacy of IR Theory as the Locus for a Pragmatist Turn', *Cooperation and Conflict*, 40/2 (2005), 147–77.

Sahm, Astrid and Westphal, Kirsten, 'Power and the Yamal Pipeline', in Margarita M. Balmaceda, James I. Clem, and Lisbeth Tarlow (eds), *Independent Belarus: Domestic Determinants, Regional Dynamics, and Implications for the West* (Cambridge, Mass: Harvard University Press, 2002).

Schiffer, Hans-Wilhelm, *Energiemarkt Bundesrepublik Deutschland*, 4. Völlig neu bearbeitete Auflage (Köln: Verlag TÜV Rheinland, 1994).

Seidelmann, Reimund, 'Die sowjetischen Energieimporte in die Bundesrepublik Deutschland, Abhängigkeit oder politische Zusammenarbeit?', *Beiträge zur Konfliktforschung*, 1 (1982), 27–66.

Simonia, Nodari, 'The West's Energy Security and the Role for Russia', *Russia in Global Affairs*, 2/3 (2004), 101–117.

Smith, David, 'Estonia: Independence and European Integration', in Artis Pabriks, Aldis Purs, Thomas Lane and David Smith (eds), *The Baltic States: Estonia, Latvia and Lithuania* (London: Routledge, 2002).

Smith, Keith C., *Russian Energy Politics in the Baltics, Poland and Ukraine: A New Stealth Imperialism?* (Washington, NY: Center for Strategic and International Studies, 2004).

Sokov, Nikolai, 'Alternative Interpretation of the Russian–Ukrainian Gas Crisis', *PONARS Policy Memo*, no. 404 (2006) <http://www.csis.org/media/csis/pubs/pm_0404.pdf>.

Soroos, Marvin S., *Beyond Sovereignty: the Challenge of Global Policy* (Columbia, SC: University of South Carolina Press, 1989).

Spero, Joan E. and Hart, Jeffrey A., *The Politics of International Economic Relations* (New York: St. Martin's Press, 1997).

Stern, Jonathan, 'The New Security Environment for European Gas: Worsening Geopolitics and Increasing Global Competition for LNG', Oxford Institute for Energy Studies, NG 15/2006.

Subacchi, Paola, 'Reforming Economic Governance in Europe: Exploring the Road to Effective Coordination', *International Affairs*, 81/4 (2005), 741–55.

Säteilyturvakeskus, *Säteilyturvakeskuksen raportti Säteilyturvallisuutta yli rajojen, Säteilyturvakeskuksen lähialueyhteistyö 1992–2002* (Helsinki: Säteilyturvakeskus, 2002).

Tassinari, Fabrizio, 'Security, Integration in the EU Neighbourhood: The Case for Regionalism', *CEPS Working Document*, no. 226 (July 2005) <http://www.ceps.be>.

Tompson, William, 'The EC's Negotiating Position with Respect to the Russian Energy Sector in the Context of Russia's WTO Accession Bid: Prospects for the Russian Federation Project' (London: The Royal Institute for International Affairs, January 2004).

Tompson, William, 'The Russian Economy under Vladimir Putin', in Cameron Ross (ed.), *Russian Politics under Putin* (Manchester: Manchester University Press, 2004).

Tsygankov, Andrei P., *Pathways after Empire: National Identity and Foreign Economic Policy in the Post-Soviet World* (Lanham: Rowman & Littlefield Publishers, 2001).

Tunander, Ola, *The Secret War Against Sweden: US and British Submarine Deception in the 1980s* (London: Frank Cass, 2004).

Ürge-Vorsatz, Diana, Miladinova, Gergana and Paizs, Laszlo, 'Energy in Transition: From the Iron Curtain to the European Union', *Energy Policy*, 34 (2006), 2279–97.

Urwin, Derek, *The Community of Europe: A History of European Integration since 1945* (London: Longman, 1995).

Vaahtoranta, Tapani and Forsberg, Tuomas, 'Post-Neutral or Pre-Allied? Finnish and Swedish Policies on the EU and NATO as Security Organisations', *UPI Working Papers*, 29/2000.

Voloshin, Vladimir, 'EU–Russia Energy Dialogue', Russian–European Centre for Economic Policy, 2004.

Wessberg, Nina, 'Mitä on suomalainen energiantuotanto?', on-line publications of the Department of Regional Studies and Environmental Policy, University Tampere <http://www.uta.fi/laitokset/yhdt/artikkelit/1999_suom_energia.pdf>.

Westphal, Kirsten, *Russische Energiepolitik: Ent- oder Neuverflechtung von Staat und Wirtschaft* (Baden-Baden: Nomos, 2000).

Westphal, Kirsten, 'Handlungsbedarf: Die Energiepolitik der Europäischen Union', *Osteuropa*, 9–10 (2004), 39–54.

Westphal, Kirsten, 'The EU-Russian Relationship and the Energy Factor: A European View', in Kirsten Westphal (ed.), *A Focus on EU-Russian Relations: Towards a Close Partnership on Defined Roadmaps?* (Frankfurt am Main et al.: Peter Lang, 2005).

Westphal, Kirsten, 'Whither Europe? Energy Policy between Multilateral Governance and Geopolitics', *Internationale Politik und Gesellschaft*, 4 (2006), 44–63.

Wiberg, Håkan, 'Adaptive Patterns and Their Deep Roots: a European Overview', in Hans Mouritzen, Ole Wæver and Håkan Wiberg (eds), *European Integration and National Adaptations: A Theoretical Inquiry* (Commack: Nova Science Publishers, 1996).

Wæver, Ole, 'The EU as a Security Actor: Reflections from a Pessimistic Constructivist on Post-Sovereign Security Orders', in Morten Kelstrup and Michael C. Williams (eds), *International Relations and the Politics of European Integration: Power, Security and Community* (London: Routledge, 2000).

Wæver, Ole, 'The Temporal Structure of European Security Identity', paper presented in the International Studies Association Annual Convention, Honolulu, HI, 2005, <http://www.isanet.org/archive.html>.

Zlobin, Nikolai, 'Limited Possibilities and Possible Limitations: Russia and the U.S.: What's Next?' *Russia in Global Affairs*, 3/1 (2005), 206–21.

Index

THE INTERNATIONAL POLITICAL ECONOMY OF NEW REGIONALISMS SERIES

Full series list

The EU–Russian Energy Dialogue
Europe's Future Energy Security
Edited by Pami Aalto

EU Development Policy and Poverty Reduction
Enhancing Effectiveness
Edited by Wil Hout

An East Asian Model for Latin American Success
The New Path
Anil Hira

European Union and New Regionalism
Regional Actors and Global Governance in a Post-Hegemonic Era
Edited by Mario Telò

Regionalism, Globalisation and International Order
Europe and Southeast Asia
Jens-Uwe Wunderlich

The Limits of Regionalism
NAFTA's Labour Accord
Robert G. Finbow

Regional Integration and Poverty
Edited by Dirk Willem te Velde and the Overseas Development Institute

Redefining the Pacific?
Regionalism Past, Present and Future
Edited by Jenny Bryant-Tokalau and Ian Frazer

Globalization and Antiglobalization
Dynamics of Change in the New World Order
Edited by Henry Veltmeyer

Latin America's Quest for Globalization
The Role of Spanish Firms
Edited by Félix E. Martín and Pablo Toral

Exchange Rate Crises in Developing Countries
The Political Role of the Banking Sector
Michael G. Hall

Persistent Permeability?
Regionalism, Localism, and Globalization in the Middle East
Edited by Bassel F. Salloukh and Rex Brynen

Reforging the Weakest Link
Global Political Economy and Post-Soviet Change in Russia, Ukraine and Belarus
Edited by Neil Robinson

Twisting Arms and Flexing Muscles
Humanitarian Intervention and Peacebuilding in Perspective
Edited by Natalie Mychajlyszyn and Timothy M. Shaw

The Political Economy of Interregional Relations
ASEAN and the EU
Alfredo C. Robles, Jr.

Asia Pacific and Human Rights
A Global Political Economy Perspective
Paul Close and David Askew

The New Political Economy of United States-Caribbean Relations
The Apparel Industry and the Politics of NAFTA Parity
Tony Heron

Transnational Democracy in Critical and Comparative Perspective
Democracy's Range Reconsidered
Edited by Bruce Morrison

Reconfigured Sovereignty
Multi-Layered Governance in the Global Age
Edited by Thomas L. Ilgen

The Nordic Regions and the European Union
Edited by Søren Dosenrode and Henrik Halkier

Japan and South Africa in a Globalising World
A Distant Mirror
Edited by Chris Alden and Katsumi Hirano

Development and Security in Southeast Asia
Volume II: The People
Edited by David B. Dewitt and Carolina G. Hernandez

The New Regionalism in Africa
Edited by J. Andrew Grant and Fredrik Söderbaum

Comparative Regional Integration
Theoretical Perspectives
Edited by Finn Laursen

The Political Economy of a Common Currency
The CFA Franc Zone Since 1945
David Stasavage

Euro-Mediterranean Security
A Search for Partnership
Sven Biscop

Development and Security in Southeast Asia
Volume I: The Environment
Edited by David B. Dewitt and Carolina G. Hernandez

Development and Security in Southeast Asia
Volume III: Globalization
Edited by David B. Dewitt and Carolina G. Hernandez

Global Turbulence
Social Activists' and State Responses to Globalization
Edited by Marjorie Griffin Cohen and Stephen McBride

New and Critical Security and Regionalism
Beyond the Nation State
Edited by James J. Hentz and Morten Bøås

Reconstituting Sovereignty
Post-Dayton Bosnia Uncovered
Rory Keane

Political Aspects of the Economic and Monetary Union
The European Challenge
Edited by Søren Dosenrode

Thailand, Indonesia and Burma in Comparative Perspective
Priyambudi Sulistiyanto

Crises of Governance in Asia and Africa
Edited by Sandra J. MacLean, Fahimul Quadir and Timothy M. Shaw